Laser shock peening

Related titles:

Surface coatings for protection against wear
(ISBN-13: 978-1-85573-767-9; ISBN-10: 1-85573-767-1)
This authoritative new book presents an overview of the current state of research in, and applications for, wear protective coatings. It concentrates on the different types of surface technologies used for wear protective coatings. Each chapter provides an in-depth analysis of a particular type of surface coating, its properties, strengths and weaknesses in various applications. Each surface coating treatment examined includes case studies describing its performance in a specific application. *Surface coatings for protection against wear* will be an invaluable reference resource for all engineers concerned with the latest developments in coatings technology.

Processes and mechanisms of welding residual stress and distortion
(ISBN-13: 978-1-85573-771-6; ISBN-10: 1-85573-771-X)
Measurement techniques for characterisation of residual stress and distortion have improved significantly. More importantly, the development and application of computational welding mechanics have been phenomenal. Through the collaboration of experts, this book provides a comprehensive treatment of the subject. It develops sufficient theoretical treatments on heat transfer, solid mechanics and materials behaviour that are essential for understanding and determining welding residual stress and distortion. The book will be useful for advanced analysis of the subject and provides examples and practical solutions for welding engineers.

Fundamentals of metallurgy
(ISBN-13: 978-1-85573-927-7; ISBN-10: 1-85573-927-5)
As product specifications become more demanding, manufacturers require steel with ever more specific functional properties. As a result, there has been a wealth of research on how those properties emerge during steelmaking. The first part of the book reviews the effects of processing on the properties of metals with a range of chapters on such phenomena as phase transformations, types of kinetic reaction, transport and interfacial phenomena. Authors discuss how these processes and the resulting properties of metals can be modelled and predicted. Part II discusses the implications of this research for improving steelmaking and steel properties. With its distinguished editor and international team of contributors, *Fundamentals of metallurgy* will be an invaluable reference for steelmakers and manufacturers requiring high-performance steels in such areas as automotive and aerospace engineering.

Laser shock peening

Performance and process simulation

K. Ding and L. Ye

Woodhead Publishing and Maney Publishing
on behalf of
The Institute of Materials, Minerals & Mining

CRC Press
Boca Raton Boston New York Washington, DC

WOODHEAD PUBLISHING LIMITED
Cambridge England

Woodhead Publishing Limited and Maney Publishing Limited on behalf of
The Institute of Materials, Minerals & Mining

Woodhead Publishing Limited, Abington Hall, Abington, Cambridge CB1 6AH, England www.woodheadpublishing.com

Published in North America by CRC Press LLC, 6000 Broken Sound Parkway, NW, Suite 300, Boca Raton, FL 33487, USA

First published 2006, Woodhead Publishing Limited and CRC Press LLC

British Library Cataloguing in Publication Data
A catalogue record for this book is available from the British Library.

Library of Congress Cataloging in Publication Data
A catalog record for this book is available from the Library of Congress.

Woodhead Publishing ISBN-13: 978-1-85573-929-1 (book)
Woodhead Publishing ISBN-10: 1-85573-929-1 (book)
Woodhead Publishing ISBN-13: 978-1-84569-109-7 (e-book)
Woodhead Publishing ISBN-10: 1-84569-109-1 (e-book)
CRC Press ISBN-10: 0-8493-3444-6
CRC Press order number: WP3444

The publishers' policy is to use permanent paper from mills that operate a sustainable forestry policy, and which has been manufactured from pulp which is processed using acid-free and elementary chlorine-free practices. Furthermore, the publishers ensure that the text paper and cover board used have met acceptable environmental accreditation standards.

Typeset by SNP Best-set Typesetter Ltd., Hong Kong
Printed by TJ International Limited, Padstow, Cornwall, England.

Contents

Preface

Laser shock peening (LSP) is an innovative surface treatment technique, which has been successfully applied to improve fatigue performance of metallic components. The key beneficial characteristic after LSP treatment is the presence of compressive residual stresses beneath the treated surface of metallic materials, mechanically produced by high magnitude shock waves induced by a high-energy laser pulse. Compared with the traditional shot peening (SP) process that has been adopted by industry for over a century to improve the surface and fatigue resistance of metallic components, LSP can produce high magnitude compressive residual stresses of more than 1 mm in depth, four times deeper than traditional SP.

LSP has been intensively investigated in the last two decades with over 100 scientific papers and reports. Most studies and investigations have been based on experimental approaches, focusing on understanding the mechanisms of LSP and its influences on mechanical behaviour and in particular enhanced fatigue performance of treated metallic components. In most cases, there has been a lack of comprehensive documentation of the relevant information in applications of LSP for various metallic alloys, such as materials properties, component geometry, laser sources, LSP parameters and the distribution of three-dimensional residual stresses. However, some comprehensive modelling capacities based on analytical models and dynamic finite element models (FEM) have been established to simulate LSP in the last decade, which provide unique tools for the evaluation of LSP and optimisation of residual stress distributions in relation to materials properties, component geometry, laser sources and LSP parameters. These approaches can play significant roles in the design and optimisation of LSP processes in practical applications.

The main aim in writing this book is to consolidate all the available knowledge and experience in a comprehensive publication for the first time. It describes the mechanisms of LSP and its significant role in improving microstructure, surface morphology, hardness, fatigue life and strength, and stress corrosion cracking. In particular, it comprehensively describes simulation techniques and procedures with some typical case studies, which can be adopted by engineers and research scientists to design, evaluate and optimise LSP processes in practical applications.

The research work from which this book arises was performed at the Centre of Expertise in Damage Mechanics (CoE-DM) supported by the Air Vehicles Division, DSTO Platforms Sciences Laboratory, Australia, from 1997 to 2003. The work was based on a research project on evaluation and characterisation of LSP for aerospace applications. The authors would like to thank their colleagues and friends for useful discussions and help in the preparation of this book. The authors are particularly grateful to Y.-W. Mai, G. Clark, C. Montross, Q. Liu, T. Wei, K. Sharp and M. Lu who contributed to the research project. Further thanks are due to F. Rose and A. Baker, whose encouragement made it possible to write this book.

Finally, L. Ye would like to thank his family, especially his wife, Pei, for their love, understanding and assistance over the years.

1
General introduction

1.1 Laser shock peening

Laser shock peening (LSP) is an innovative surface treatment technique, which is successfully applied to improve fatigue performance of metallic components. After the treatment, the fatigue strength and fatigue life of a metallic material can be increased remarkably owing to the presence of compressive residual stresses in the material. The increase in hardness and yield strength of metallic materials is attributed to high density arrays of dislocations (Banas *et al.*, 1990a, b) and formation of other phases or twins (Chu *et al.*, 1999), generated by the shock wave.

The ability of a high energy laser pulse to generate shock waves and plastic deformation in metallic materials was first recognised and explored in 1963 in the USA (White, 1963). The confined ablation mode for an improved LSP process was established in 1968 (Anderholm, 1970). The LSP process was initially performed to investigate its application for the fastener holes in 1968–1981 at Battelle Columbus Laboratories (OH, USA) (Clauer *et al.*, 1981). Since 1986, more systematic studies on applications of LSP have been carried out in other countries, such as France (Ballard *et al.*, 1991; Peyre and Fabbro, 1995a, b), China (Zhang and Cai, 1996; Dai *et al.*, 1997; Guo *et al.*, 1999) and Japan (Sano *et al.*, 1997).

Since the development of LSP, a number of patents have been issued addressing its strong interest for commercialisation. In 1974, the first patent was issued after the benefits of LSP were clearly identified (Mallozi and Fairand, 1974). For example, laser peening of braze repaired turbine components (Mannava and Ferrigno, 1997; Mannava *et al.*, 1997) and weld repaired turbine components (Ferrigno *et al.*, 1998) have been patented because of the clear improvement in properties.

Although the conventional shot peening (SP) treatments have existed in industry for over six decades, the LSP process, producing impressive compressive residual stresses into metallic materials, is envisaged as a substitute for SP conventional treatments to improve the fatigue performance of those materials. The increased in-depth compressive residual stresses produced by LSP can significantly improve fatigue performance of materials, strengthening thin sections and controlling development and growth of surface cracks (Dane *et al.*, 1997; Mannava and Cowie, 1996).

An LSP process can be used to treat various kinds of metallic components, which include cast irons, aluminium alloys, titanium and its alloys, nickel-based superalloys and so forth. In the aerospace industry, LSP can be used to treat many aerospace products, such as turbine blades and rotor components (Mannava and Ferrigno, 1997; Mannava *et al.*, 1997), discs, gear shafts (Ferrigno *et al.*, 2001) and bearing components (Casarcia *et al.*, 1996). In particular, LSP has clear advantages for treating components of complex geometry such as fastener holes in aircraft skins and refurbishing fastener holes in old aircraft, where the possible initiation of cracks may not be discernible by normal inspection. Protection of turbine engine components against foreign object damage (FOD) is a key concern of the US Air Force (Zhang *et al.*, 1997). General Electric Aircraft Engines treated the leading edges of turbine fan blades (Mannava *et al.*, 1997) in an F101-GE-102 turbine using LSP for the Rockwell B-1B bomber, which enhanced fan blade durability and resistance FOD without harming the surface finish. In addition, it was reported that LSP would be applied to treat engines used in the Lockheed Martin F-16C/D (Obata *et al.*, 1999). The laser peened components, which can significantly enhance the resistance to fatigue, fretting, galling and stress corrosion are well appreciated by the research community (Banas *et al.*, 1990a, b; Chu *et al.*, 1995; Peyre *et al.*, 1995; Clauer, 1996; Dane *et al.*, 1997).

A laser pulse that can be adjusted and controlled in real time is a unique advantage of LSP (Mannava, 1998). Through a computer controlled system, the energy per pulse can be measured and recorded for each LSP process on the component. In particular, multiple LSP can be applied at the same location. Regions inaccessible by shot peening (SP), such as small fillets and notches, can still be treated by LSP (Mannava and Cowie, 1996).

A schematic configuration of an LSP process on a metal plate is shown in Fig. 1.1. When shooting an intense laser beam on to a metal surface for a very short period of time (around 30 ns), the heated zone is vaporised to reach temperatures in excess of 10 000 °C and then is transformed to plasma by ionisation. The plasma continues to absorb the laser energy until the end of the deposition time. The pressure generated by the plasma is transmitted to the material through shock waves. The interaction of the plasma with a metal surface without coating is defined as 'direct ablation', which can achieve a plasma pressure of some tenths of a GPa (Sano *et al.*, 1997; Masse and Barreau, 1995a, b). In order to obtain a high amplitude of shock pressure, an LSP process normally uses a confined mode, in which the metal surface is usually coated with an opaque material such as black paint or

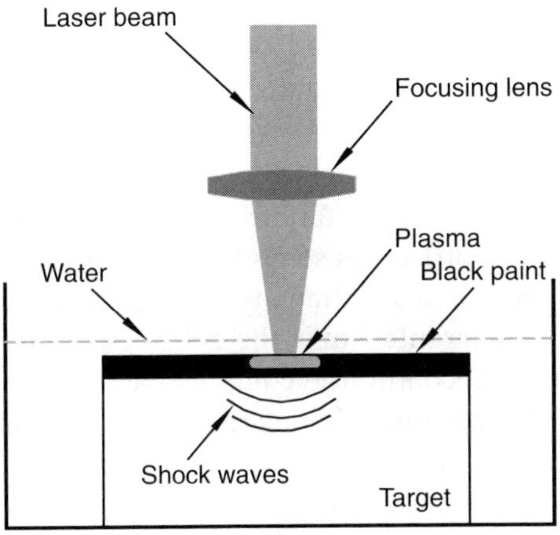

1.1 Schematic configuration of laser shock peening.

aluminium foil, confined by a transparent material such as distilled water or glass against the laser radiation. This type of interaction is called 'confined ablation'. Recent research had found that, when using the confined mode, ever greater plasma pressures of up to 5–10 GPa could be generated on the metal surface (Fairand *et al.*, 1974; Devaux *et al.*, 1991; Berthe *et al.*, 1997; Bolger *et al.*, 1999). A stronger pressure pulse may enhance the outcome of LSP with a high magnitude of compressive residual stress to a deeper depth.

The laser spot size and geometry of LSP can be tailored for individual applications and a laser spot with either a square profile or a round one has been used in practice. Furthermore, the LSP process is clean and workpiece surface quality is essentially unaffected, especially for steel components. LSP also has the potential to be directly integrated into manufacturing production lines with a high degree of automation (Mannava, 1998).

The applications of LSP can be anticipated to expand from the current field of high value, low volume parts such as aircraft components to higher volume ones such as the automobile, industrial equipment and tooling in the near future as high power, Q-switched laser systems become more available (Clauer, 1996).

1.2 Traditional shot peening

A traditional surface treatment technique, shot peening (SP), has been effectively and widely applied in industry for over six decades. In an SP process, metal or ceramic balls acting as a minuscule ball-peen hammers make a small indentation or dimple on the metal surface on impact. A compacted volume of highly shocked and compressed material can be produced below the dimple. A thin uniform layer provided by overlapping dimples can be extremely resistant to initiation and propagation of cracks as well as corrosion, protecting the peened area (Khabou *et al.*, 1990; Li *et al.*, 1991; Thompson *et al.*, 1997). The advantages of SP are that it is relatively inexpensive, using robust process equipment and it can be used on large or small areas as required. But it also has its limitations. Firstly, in producing the compressive residual stress, the process is semi-quantitative and is dependent on a metal strip or gauge called an Almen type gauge to define the SP intensity. This gauge cannot guarantee that the SP intensity is uniform across the component surface. Secondly, the compressive residual stress is limited in depth, usually not exceeding 0.25 mm in soft metals such as aluminium alloys and less in harder metals (Clauer, 1996). Thirdly, the process results in a roughened surface, especially in soft metals like aluminium. This roughness may need to be removed for some applications, while typical removal processes often resulted in the removal of the majority of the peened layer.

In comparison, an LSP process can produce a compressive residual stress more than 1 mm in depth, which is about four times deeper than the traditional SP process (Clauer, 1996). In addition, an SP process may damage the surface finish of metal components and can easily cause distortion of thin sections, whilst in LSP, the treated surface of the component is essentially unaffected and the laser peened parts do not lose any dimensional accuracy normally. The LSP process is a better and more effective way to achieve the same outcome with less disadvantages. Moreover, as the laser pulse can be adjusted and optimised, the process can become more efficient in application. Despite the fact that the use of the LSP process is much more expensive than the SP process, some manufacturers still endeavour to use LSP to treat some critical metal components such as engine blades for aircrafts.

1.3 Scope of the book

The improved properties and microstructural changes in metallic materials induced by LSP have widely been recognised by many researchers (Fairand *et al.*, 1972; Clauer *et al.*, 1977; Banas *et al.*, 1990b; Chu *et al.*, 1995; Peyre *et al.*, 2000a). Since the mid-1980s researchers have conducted many experiments to elaborate the effects of the confined interaction mode and the factors influencing the laser pulse during an LSP process.

The main function of LSP is to introduce surface compressive residual stress or surface strain hardening that can lead to an improvement in the mechanical performance of metallic components such as fatigue and corrosion resistance. The distribution of residual stress in a peened metallic component is clearly dependent on the generation and propagation of a shock wave (or dynamic stress) and its interaction with the component, for example the geometry and material properties in relation to a single or multiple LSP process. Typical LSP parameters for a confined ablation mode include the laser power density, deposition time and laser spot size. It is appreciated that inappropriate combination of these factors for an LSP process can induce significant tensile residual stresses that can be very detrimental to the mechanical performance of the component.

However, the use of experimental instruments to characterise the shock wave or dynamic stress in a laser peened component can be very expensive and complicated. For a better understanding of LSP, and in order to optimise its process by addressing the various factors mentioned above, simulation based on mechanistic modelling using analytical or finite element methods has currently been recognised as an effective tool in the approach, if the simulation procedures have been well calibrated and validated by the experimental data. The aim of this work is to present the state-of-the-art of LSP in terms of its mechanisms, performance and process simulation. In terms of process simulation, it will focus on the knowledge and experience of the authors in using finite element modelling (FEM) in simulating LSP on metal components of different geometry. Dynamic stresses and residual stresses in laser peened metallic components are investigated. Some influential parameters associated with LSP are evaluated for the purpose of characterising LSP processes. In particular, different methods of using LSP, such as one-sided, two-sided and multiple LSP on flat or curved surfaces of components are elaborated in detail. The outline of this book is described as follows.

- Chapter 2 presents a comprehensive literature review of the physical and mechanical mechanisms of LSP for metallic materials, which have been investigated in the past 30 years. In particular, attention has been focused on physical models of LSP with key parameters. The effects of LSP on mechanical properties of metallic alloys are also highlighted.
- Chapter 3 introduces the simulation methodology of LSP, addressing procedures of both analytical modelling and FEM simulation. Especially, some important algorithms involved in simulation of LSP are highlighted.
- Chapter 4 presents simulations of single and multiple LSP with a round laser spot on a flat surface using a two-dimensional (2D) finite element model. The effects of mesh refinement, bulk viscosity, material damping and some other influential parameters of LSP are elaborated. Correlation between predicted results and experimental data is also evaluated.
- Chapter 5 describes simulations of single and multiple LSP with a square laser spot on a flat surface using a three-dimensional (3D) finite

element model. Further studies of dynamic behaviour, compressive residual stress and plastically affected depth are presented and discussed. Simulated results are correlated with experimental data.

- Chapter 6 presents simulations of single and multiple LSP on opposite surfaces of thin flat sections using 2D finite element models. The results are carefully evaluated and discussed with respect to changes in some influential factors related to LSP. Predictions are compared with available experimental data.
- Chapter 7 presents simulations of single and multiple LSP on opposite surfaces of bar specimens of a circular cross-section using both 2D and 3D finite element models. The emphasis is placed on evaluating potential harmful tensile residual stresses at the middle of the cross-section with respect to changes in some influential factors related to LSP. The effects of residual tensile stresses are correlated to experimental data.

2

Physical and mechanical mechanisms of laser shock peening

2.1 Introduction

After laser shock peening (LSP) was invented in the early 1960s, the studies mainly focused on the basic process development, understanding of mechanisms, the use of high laser power density to achieve high pulse pressures (Fairand *et al.*, 1972) and development of physical models to characterise LSP processes (Peyre *et al.*, 1996). Since 1986, many researchers (Ballard *et al.*, 1991; Devaux *et al.*, 1991; Peyre and Fabbro, 1995a, b; Peyre *et al.*, 1995; Berthe *et al.*, 1997) have further developed and enriched this technique by addressing effects of modified laser temporal shape, characteristics of shock waves and their propagation as well as modelling the induced mechanical responses. Much attention in the studies was paid to some influential factors related to LSP conditions, such as laser parameters, confined overlays and thermoprotected coatings, which can significantly affect the mechanical responses of the metallic materials.

This chapter presents an overview of the state of the art of LSP, highlighting its physical mechanisms and its effect on the mechanical performance of treated metallic components. Emphasis is placed on essential aspects of LSP including laser power density, pulse shape and duration, pulse rise time, laser wavelength, laser spot, thermal protective coating and confining overlay to conserve the plasma energy, as well as multiple shots and the coverage ratio of impacts.

2.2 Laser systems for laser shock peening

In order to fulfil the LSP process requirements, it is very important to select a suitable laser system, which normally requires an average power level of from several hundreds watts to kilowatts, a pulse energy of around 100 J and a pulse duration of around 30 ns. In addition, both a high repetition rate of the laser pulse and a reasonable laser wavelength are also important for LSP to assure effective treatment results for metal components. Selecting a laser system for LSP application not only requires these physical characteristics of the laser source, but also needs to consider some specific requirements such as cost, efficiency, maintenance and part replacements and so on.

The neodymium-doped glass (Nd-glass) laser was initially developed in 1974 at Battelle Columbus Laboratories (BCL) Ohio. It was quite cumbersome, about 150 ft long (though powerful, >500 J per pulse), and its repetition rate was extremely slow, about one cycle every 8 min. Later, based on this technology, BCL sponsored by Wagner Laser Technologies (WLT) invented a 4 × 6 ft (~1.2 × 1.8 m) glass-laser system capable of 100 J or so, with a repetition rate of 1 Hz, or one cycle per second (cps) (Vaccari, 1992). The Lawrence Livermore National Laboratory (LLNL) has continuously developed a high power Nd-glass laser systems for fusion applications over the past 25 years (Dane *et al.*, 1997). One of the laser systems can deliver an average pulse energy of 25–100 J, repetition rates of up to 10 Hz and an average power level near 1 kW.

In most LSP processes, laser beams are produced by a Q-switched laser system based on a neodymium-doped glass or yttrium aluminium garnet (YAG) crystal lasing rod, which operates in the near infrared, having a wavelength of 1.064 μm and a pulse duration of 10–100 ns. Table 2.1 shows some typical laser systems with reported processing parameters for LSP in the open literature. In general, development of the laser systems is very important for successful industrial applications of LSP. A suitable system should have an energy output in the range of 10–500 J/pulse with a pulse duration of less than 100 ns. The wavelength of the laser is also a very important parameter because it controls the interaction between the laser beam and the material surface. In the near future, laser systems of much better output performance may be achieved by using advanced technology such as diode pumped and slab technology and these can greatly facilitate diffusion of LSP and broaden its industrial applications (Fabbro *et al.*, 1998).

2.3 Generation of a shock wave

With the invention of the laser, it was soon recognised that the high amplitude of shock waves required for a SP process could be achieved by using confined plasma generated at the metal surface by means of a high-intensity laser beam with a pulse duration in the tens of nanoseconds range (Dane *et al.*, 1997).

The physics and mechanisms of laser-induced shock wave generation has been investigated intensively (Fairand *et al.*, 1974; O'Keefe and Skeen, 1972; Hoffman, 1974; Yang, 1974; Romain *et al.*, 1986; Ling and Wight, 1995; Couturier *et al.*, 1996). In early experiments (White, 1963; Skeen and York, 1968), the irradiated material was placed in a vacuum and the plasma

Table 2.1 Typical laser systems used for LSP processes

Treated materials	Laser type	Laser power (J)	Power density (GW/cm^2)	Pulse duration (ns)	Laser spot size (mm)	Absorbent coating	Transparent overlay	Peak pressure (GPa)	References
Rock	Nd: glass	5–100	1–15	20	2–6.6	–	Water	1.4	Bolger *et al.*, 1999
Al foil	Nd: glass	40	0–25	25–30	3–5	–	Water (glass)	5.5	Berthe *et al.*, 1997
Thin Al	Nd: YAG	6	0.05–1	150	3	–	–	0.8	Griffin *et al.*, 1986
Al 2024-T351 and T851, 7075-T631 and T73	Nd: glass	–	–	20–30	0.6–3	Black paint	Water (quartz)	10	Clauer and Fairand, 1979
2024-T3 Al	Nd: glass	–	5	18	10	Black paint	Water	–	Yang *et al.*, 2001
2024-T62 Al	Nd: glass	–	1.57–7.32	18–23	6–8	Black paint	K7 glass	–	Zhang and Yu, 1998
Al-12Si, A356 Al, 7075Al	Nd: glass	80	1–8	15–30	5–12	Black paint	Water	2.5	Peyre *et al.*, 1996
Ti-6Al-4V	Nd: glass	–	5.5–9	–	5.6	Black paint	Water	–	Smith *et al.*, 2000
SUS304 s.	Nd: YAG	0.1	4.5	5	0.75	–	Water	0.5	Sano *et al.*, 1997
316L s.s.	Nd: glass	40–100	8–20	3–10	–	Black paint	Water	10	Peyre *et al.*, 2000b
Al, 55C1 s., 316L s.s.	Nd: glass	40	8–10	8–10	3–4	Al paint	Water	6	Peyre *et al.*, 1998a

Table 2.1 Continued

Treated materials	Laser type	Laser power (J)	Power density (GW/cm^2)	Pulse duration (ns)	Laser spot size (mm)	Absorbent coating	Transparent overlay	Peak pressure (GPa)	References
316L s.s., X12CrNi12-2-2 s.	Nd: glass	40–100	1–100	0.6–30	0.5–1	Al foil, Al adhesive	Water	6	Peyre *et al.*, 1998b
Hypoeutectoid s.	Nd: glass	80	5–10	25	5	Black paint	Water (BK7 glass)	5	Masse and Barreau, 1995a, b
Fe-30%Ni Al	KDP	4000	10^2–10^4	1	4.3–25	–	–	0.6	Grevey *et al.*, 1992
304 s.s.	Nd: glass	80	300	0.6	7.2	Black paint	Water	18	Gerland *et al.*, 1992
Hadfield manganese	Nd: glass	100	2400	0.6	3–3.5	Black paint	Quartz	39.5	Chu *et al.*, 1995
18Ni(250) s.	Nd: YAG	0.03	1000	0.15	0.1	Black paint	Water	–	Banas *et al.*, 1990b

s. = steel, s.s. = stainless steel.
KDP = potassium dihydrogen phosphate.

generated by the laser pulse expanded freely. The resulting peak plasma pressure ranged from 1 GPa up to 1 TPa when the laser power density was varied from about 0.1 GW/cm^2 to 10^6 GW/cm^2. The time duration of the plasma pressure was roughly equal to the laser pulse duration, typically 50 ns in length, because of the rapid adiabatic cooling of the laser-generated plasma in the vacuum (Fairand and Clauer, 1978; Clauer *et al.*, 1981).

There are three wavelengths use of most commonly in LSP processes, 1.064 μm (near infrared, IR), 0.532 μm (green) or 0.355 μm (ultraviolet, UV). The near infrared wavelength has only a modest absorption coefficient in a water overlay, sufficient interaction with the metal surface and a high dielectric breakdown threshold, while the green wavelength has the lowest absorption in a water overlay. Berthe *et al.* (1999) first conducted studies into the characterisation of laser shock waves and the effects of the breakdown of plasma with respect to laser wavelengths emitting from IR to UV laser sources. The results indicated that, when the laser power density was increased, the pressure produced by a laser pulse with wavelengths in the green and UV had a similar profile to that generated with a wavelength in the IR. In addition, the pressure produced by a laser pulse with a wavelength in the IR, corresponding to a laser power density of 10 GW/cm^2, was saturated at 5.0 GPa with the water-confined mode (WCM). But saturated pressure at UV and green wavelengths occurred at higher laser power densities than at IR wavelength. Moreover, the pressure durations with UV wavelength decreased more strongly than with IR wavelength. Therefore, the breakdown plasma in a WCM was favoured by shorter wavelengths.

Although metals can be highly reflective of light, keeping the constant laser power density and decreasing the wavelength from IR to UV can increase the photon–metal interaction enhancing shock wave generation. However, the peak plasma pressure may decrease because decreasing the wavelength decreases the critical power density threshold for a dielectric breakdown, which in turn limits the peak plasma pressure (Fairand *et al.*, 1974; Berthe *et al.*, 1999). The dielectric breakdown is the generation of plasma not on the material surface, which absorbs the incoming laser pulse, limiting the energy to generate a shock wave. In Fig. 2.1, the decrease in the wavelength from IR to green reduces the dielectric breakdown threshold from 10–6 GW/cm^2, resulting in maximum peak pressures of approximately 5.5 and 4.5 GPa, respectively.

Berthe *et al.* (1997) studied parasitic plasma and pressure measurement in LSP processes with a WCM using of two types of instruments, the velocimetry interferometer system for any reflector (VISAR) and a fast camera. They found that the experimental measured pressure was a function of laser power density. The experimental data associated with the relationship between the pressure and laser power density reveals that, when

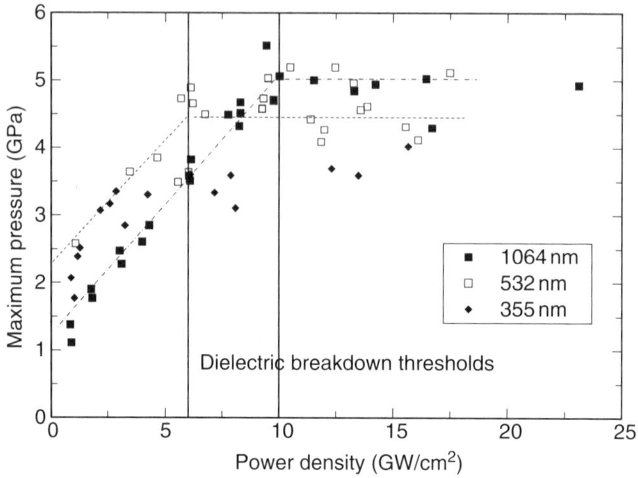

2.1 Peak plasma pressures obtained in WCM as a function of laser power density at 1.064 μm (Berthe *et al.*, 1997), 0.532 μm and 0.355 μm laser wavelength (Berthe *et al.*, 1999).

increasing the laser power density from 1 to 10 GW/cm^2, the pressure is also increased; but when the laser power density is increased above 10 GW/cm^2, the corresponding pressure is scattered and saturated. The saturation of the pressure is attributed to the confining water breakdown phenomenon that limits the laser power density reaching material surface.

Other researchers, such as Fabbro *et al.* (1990), Devaux *et al.* (1993) and Sollier *et al.* (2001), also discussed the confining water breakdown phenomenon. The breakdown phenomenon has two detrimental effects on the shock waves induced into the material when increasing the laser power density above 10 GW/cm^2: (1) the peak pressure is saturated; (2) the pressure duration is shortened (Berthe *et al.*, 1997).

In the LSP process with a WCM, the saturation of the peak pressure can reach as high as 5.5 GPa with a duration of 55 ns. In treating many high strength metallic materials, these LSP conditions are very useful for a deep treatment (Devaux *et al.*, 1993; Berthe *et al.*, 1997). However, if the laser power densities are less than 0.1 GW/cm^2, no shock waves can be created within the material. In addition, if the laser power densities are around 1 GW/cm^2, the shock wave formation is unaffected by material thermal properties (Clauer *et al.*, 1981).

A suitable laser system can produce a high-energy laser pulse to offer an ideal source for LSP. If laser parameters, such as the laser power density (I_0), laser spot size (D) and laser pulse duration (τ), were optimised appropriately, the optimised process could improve the mechanical properties and microstructures of the metal alloy components enormously. Zhang and

Yu (1998) studied optimisation of the laser parameters to improve LSP processes on the metallic materials. They found that the laser power density in a range between $[64(\sigma_Y^{dyn})^2/MZA]$ and $[64(\sigma_U^{dyn})^2/MZA]$ for the LSP process produced a better treatment result. In this expression, A is the absorption coefficient of the surface coating, M is the transmission coefficient of the transparent overlay, Z is the reduced shock impedance between the metal and the transparent overlay, σ_Y^{dyn} is the dynamic yield strength of the metal and σ_U^{dyn} is the dynamic ultimate tensile strength of the metal.

The use of laser-absorbent sacrificial coatings has also been found to increase the shock wave intensity in addition to protecting the metal surface from laser ablation and melting. Metal coatings such as aluminium, zinc or copper and organic coatings have been found to be beneficial if not necessary to protect the component surface (Fairand *et al.*, 1974). Among the absorbent coatings, commercially available flat black paint has been found to be practical and effective, compared to other coating systems (Montross and Florea, 2001).

It was observed that the use of transparent overlays, such as water or glass, with the laser energy could increase the shock wave intensity propagating into the metal by up to two orders of magnitude, as compared to plasmas generated in a vacuum state (Fairand *et al.*, 1972; Fairand *et al.*, 1974; Fabbro *et al.*, 1990). Because the transparent overlay prevents the laser-generated plasma from expanding rapidly away from the surface, an increase in shock wave intensity can be achieved. The transparent overlay results in more of the laser energy being delivered into the material as a shock wave than without it (Montross *et al.*, 1999).

When a laser pulse with sufficient intensity irradiates a metal material with an absorbent coating through the transparent overlay, the absorbent material vaporises and forms high-energy plasma. Because of the short period of energy deposition, the diffusion of thermal energy away from the interaction zone is limited to a couple of micrometres and should preferably be less than the thickness of the absorbent coating to maintain protection. It is critical for aluminium alloys since surface ablation processes can affect fatigue life detrimentally (Fairand and Clauer, 1977). The plasma continues to absorb the laser energy until the end of the energy deposition (Fairand and Clauer, 1978).

The hydrodynamic expansion of the heated plasma in the confined region between the metal material and the transparent overlay creates a high amplitude, short duration, pressure pulse. As a result, shock waves are created, propagating into the metal. When the stress of the shock wave exceeds the dynamic yield strength of the metal, plastic deformation occurs, which consequently modifies the near-surface microstructure and properties (Clauer, 1996).

2.4 Measurement of residual stress

Residual stresses after LSP are the stresses remaining in a metal after the shock wave is dispersed. Such residual stresses play a key role in enhancing the fatigue performance of metallic materials. The measurement of residual stresses allows engineers to understand fully the residual stress profile in the treated metallic components. Thus, an accurate residual stress measurement is important in the design and quality control of mechanical or thermal treatment processes for metal components. The residual stress is often measured using a special technique such as centre-hole drilling, layer removal, X-ray diffraction or neutron diffraction and so on (Lu, 1996). The main technical characteristics of these method are described as follows.

One of the most widely used techniques for measuring residual stress is the hole-drilling strain gauge method. The general principle of the procedure involves drilling a small hole into a specimen containing residual stresses. A special residual stress strain gauge rosette, allowing back calculation of residual stress to be made, can measure the relieved surface strains. This method is semi-destructive and cannot be checked by repeat measurement.

The layer removal technique is often used for measuring the presence of residual stress in simple test piece components. The methods are generally quick and require only simple calculations to relate the curvature to the residual stresses. When layers are removed from one side of a flat plate containing residual stresses, the stresses become unbalanced, leading to bend of the plate. The curvature depends on the original stress distribution present in the layer that has been removed and on the elastic properties of the remainder of the plate. By carrying out a series of curvature measurements after successive layer removals, the distribution of stress in the original plate can then be deduced.

X-ray diffraction is a common non-destructive testing (NDT) technique that can be used to determine the levels of residual stress in a component. X-rays probe a very thin surface layer of material (typically tens of micrometres). This method is based on the use of lattice spacing as the strain gauge (Prevéy, 1996). Through knowledge of the wavelength, the change in the Bragg angle and the changes in interplanar spacings, the elastic strain may be calculated. The residual stress gradients in metallic components have typically been measured using X-ray diffraction with destructive etch/layer removal.

Synchrotrons, or hard X-rays, provide very intense beams of high energy X-rays. These X-rays have a much higher depth penetration than conventional X-rays, around 1–2 mm in many materials. This increased penetration

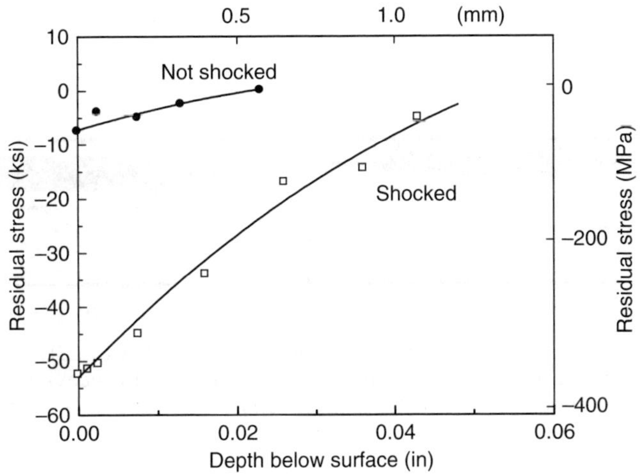

2.2 The magnitude depth of residual stresses in 6 mm thick 2034-T3 aluminium alloy (Clauer and Koucky, 1991).

depth means that synchrotron diffraction is capable of providing high spatial resolution, 3D maps of strain to millimetre depths in engineered components. This increased penetration depth is one of the major advantages of synchrotron diffraction over conventional X-ray diffraction.

Like the X-ray diffraction technique, neutron diffraction relies on elastic deformations within a polycrystalline material that cause changes in the spacing of the lattice planes from their stress-free value. Measurements are carried out in much the same way as with X-ray diffraction, with a detector moving around the sample, locating the positions of high intensity diffracted beams.

In the past 30 years, researchers have focused on the experimental investigations of surface and in-depth residual stresses induced in different LSP configurations for a number of industrial metals, such as aluminium alloys (Clauer *et al.*, 1981; Zhang and Lu, 1998; Clauer *et al.*, 1992), steels (Grevey *et al.*, 1992; Banas *et al.*, 1990a, b) and titanium alloys (Ruschau *et al.*, 1999). Most measurements of residual stresses were performed using two methods, X-ray diffraction and the centre-hole drilling technique. It was observed that the distribution of the compressive residual stress across the treated area is relatively uniform after a typical LSP treatment. The residual stress is usually highest at the surface and decreases gradually with distance below the surface. Figure 2.2 gives a typical profile for the residual stress in the depth of a 2024-T3 aluminium alloy, showing that the compressive stresses reach a depth of over 1 mm (Clauer and Koucky, 1991).

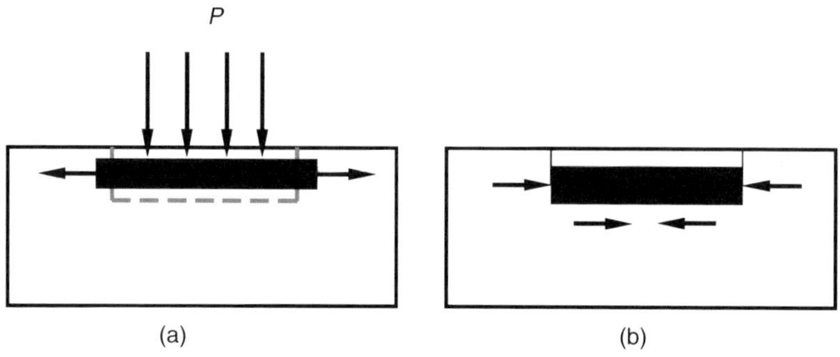

2.3 Generation of compressive residual stresses with LSP. (a) Stretching of impact area during the interaction, (b) recovery of surrounding material after laser pulse is switched off (Peyre *et al.*, 1996).

2.5 Characteristics of residual stresses induced by laser shock peening

2.5.1 Physical models of residual stress

When the laser power density reaches a level of several GW/cm^2, high-amplitude shock waves, through rapid expansion of high-temperature (around 10 000 °C) plasma of a pressure of about several GPa, can be generated in the metallic component. In a confined ablation mode, the laser energy is deposited on the plasma between the material and the transparent overlay, which continues to be heated, vaporised and ionised. As the plasma is trapped between the material and transparent overlay, the magnitude and duration of plasma can be increased by a factor of 10 for the peak pressure and by a factor of 3 for the duration, respectively, compared with the direct ablation mode (Peyre *et al.*, 1998a). Based on this confined ablation mode, an LSP process may be described by a two-step sequence: (1) the rapid plasma expansion creates sudden uniaxial compression on the irradiated area and dilation of the surface layer and (2) the surrounding material reacts to the deformed area, generating a compressive stress field (Peyre and Fabbro, 1995b; Fabbro *et al.*, 1998), shown in Fig. 2.3.

During LSP, the pressure pulse generated by the blow-off of the plasma impacts on the treated area and creates almost pure uniaxial compression in the direction of the shock wave propagation and tensile extension in the plane parallel to the surface. After the reaction in the surrounding zones, a compressive stress field is generated within the affected volume, while the underlying layers are in a tensile state (Peyre and Fabbro, 1995b; Peyre *et al.*, 1995). As the shock wave propagates into the material, plastic deformation occurs to a depth at which the peak stress no longer exceeds the Hugoniot elastic limit (HEL) of the material, which induces residual stresses throughout the affected depth. HEL is related to the dynamic yield strength according to (Johnson and Rohde, 1971):

$$\mathrm{HEL} = \frac{(1-\nu)\sigma_{\mathrm{Y}}^{\mathrm{dyn}}}{(1-2\nu)} \qquad [2.1]$$

where ν is Poisson's ratio and $\sigma_{\mathrm{y}}^{\mathrm{dyn}}$ is the dynamic yield strength at high strain rates.

When the dynamic stresses of shock waves within a material are above the dynamic yield strength of the material, plastic deformations occurs, which continues until the peak dynamic stress falls below the dynamic yield strength. The plastic deformation induced by the shock waves results in strain hardening and compressive residual stresses at the material surface (Ballard *et al.*, 1995; Peyre and Fabbro, 1995b; Dai *et al.*, 1997).

Knowledge of the plasma pressure (spatially and temporally) at the interface between the material and the transparent overlay is of primary importance for the control and optimisation of LSP (Fairand *et al.*, 1974; Fabbro *et al.*, 1990; Devaux *et al.*, 1991). There are several techniques for estimating the plasma pressure, such as using a piezoelectric quartz gauge (Anderholm, 1970; Devaux *et al.*, 1993), a piezoelectric copolymer (Couturier *et al.*, 1996) and a VISAR device (Berthe *et al.*, 1997; Peyre *et al.*, 1998a).

Fabbro *et al.* (1990) initially performed a physical and mechanical study of the laser-induced plasma to estimate the plasma pressure. Their model is based on the physical and mechanical behaviour of the laser-induced plasma, describing an LSP process in three steps. In the first step, a laser pulse irradiates the material with the transparent overlay, creating expansion of confined plasma of high pressure that drives shock waves into the material. The second step begins after switching off the laser pulse, and the plasma is characterised by adiabatic cooling, but maintains the pressure over a period twice as long as the laser pulse duration. The third step is associated with the further adiabatic cooling of the plasma, but during this period the exerted pressure was too low to drive the shock waves further into the material. The laser and pressure pulses, monitored with a fast photodiode and an x-cut quartz gauge system, respectively, are illustrated in Fig. 2.4.

Using such a model, and considering the plasma to be a perfect gas, the scaling law of peak plasma pressure, P, can be expressed as (Fabbro *et al.*, 1990):

$$P(\mathrm{GPa}) = 0.01\sqrt{\frac{\alpha}{2\alpha+3}}\sqrt{Z(\mathrm{g/cm^2\,s^2})}\sqrt{I_0(\mathrm{GW/cm^2})} \qquad [2.2]$$

where I_0 is the laser power density, α is the efficiency of the interaction and Z is the reduced shock impedance between the material and the confining

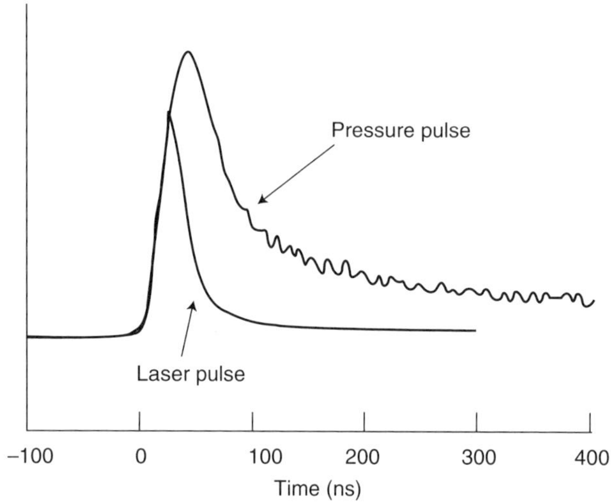

2.4 Gaussian laser pulse and resulting pressure pulse on a target (Peyre *et al.*, 1996).

medium. In a water-confined ablation mode, the peak pressure is approximately the square root of the incident laser power density.

The basic mechanics of the shock wave and the induced plastic deformation with resulting residual stresses are difficult to characterise analytically because of the three-dimensional nature of the dynamic stress state. Most explosive work is normally assumed to generate large planar shock waves, which can be simplified and analysed in a one-dimensional state. An early analysis of shock wave propagation was attempted using hydrodynamic shock wave codes and the predicted results crudely matched the experimental results (Clauer *et al.*, 1977). In line with the analyses of explosive-driven shock waves, high power lasers were used to cause spalling of aluminium and copper foils. These experimental data were compared with the results from various one- and two-dimensional analytical computer codes with reasonable agreement (Cottet *et al.*, 1988; Cottet and Boustie, 1989; de Rességuier *et al.*, 1997).

Ballard *et al.* in 1991 established the first analytical model for residual stress field in a material after LSP. Based on the mechanical behaviour of the material induced by a pulse pressure, Ballard (1991) assumed that the material is a semi-infinite body with some assumptions to estimate the plastically affected depth and the peak compressive residual stress in the material. Peyre *et al.* (1996) first applied the model to correlate with their experimental data on LSP of aluminium alloys.

The plastic deformation in the material depends on the HEL (Peyre *et al.*, 1998b). During LSP, if the peak dynamic stress is below HEL, no plastic deformation occurs in the material. If the peak dynamic stress is between 1 and 2 HEL, the plastic strain occurs with a purely elastic reverse strain. If the peak dynamic stress is above 2 HEL, the elastic reverse strain gets saturated and the plastic strain fully occurs (Peyre and Fabbro, 1995b; Fabbro *et al.*, 1998). Beyond P = 2 HEL, no further plastic deformation occurs. Therefore, materials are optimally treated with a peak dynamic stress in the 2–2.5 HEL range so that a maximum surface plastic strain can be obtained in the material (Ballard *et al.*, 1991; Peyre and Fabbro, 1995b).

2.5.2 Transparent overlay and absorbent coating

In LSP without transparent overlay, the laser-induced plasma absorbs the incident laser energy and it expands freely from the solid surface. Consequently, the incident laser energy cannot efficiently be converted into a pressure pulse that induces compressive residual stresses in the substrate by a shock wave.

The transparent overlay can be any transparent materials, such as water, glass, fused quartz and acrylic, which are used as a confined overlay in LSP. The confined overlay can trap thermally expanding plasma over the metal surface, causing the plasma pressure to rise much higher than it would be if the transparent material were absent (Fairand *et al.*, 1974; Clauer and Fairand, 1979; Masse and Barreau, 1995a, b; Bolger *et al.*, 1999). The confined overlay is normally placed over the thermal protective material coated on the material surface. To be a confined overlay, the simplest and most cost-effective material is a thin water layer flowing over the coated metal surface from an appropriately placed nozzle.

Sano *et al.* (1997) conducted an experiment to observe laser-induced plasma generated by the SH-YAG laser with the direct ablation mode or the WCM. It was observed that the plasma pressure was significantly increased by the presence of the water confinement, compared with that of the direct ablation mode. Hong *et al.* (1998) later studied characteristics of laser-induced shock waves under five kinds of confined overlays including Perspex, silicon rubber, K9 glass, quartz glass and Pb glass. The experiments were performed with an Nd:glass laser. The material was a 2024T62 aluminium alloy coated with a black paint. The measured peak pressure from these five confined overlays is shown in Table 2.2. The peak pressure can be increased when selecting a confined overlay of high acoustic impedance and meanwhile the pressure duration can also be significantly widened using an overlay of high acoustic impedance. However, for overlays with low acoustic impedance, the pressure duration is nearly equal to the laser pulse duration (40 ns).

Table 2.2 Peak pressures with five confined overlays (Hong *et al.*, 1998)

Confined overlay	Acoustic impedance Z (10^6 g/cm^2 s)	Laser power density I_0 (10^9 W/cm^2)	Pressure duration τ (ns)	Experimental results P_{max} (10^8 Pa)
Perspex	0.32	0.74	53	11.3
Silicon rubber	0.47	0.74	54	13.8
K9 glass	1.14	0.68	160	15.9
Quartz glass	1.31	0.76	131	17.2
Pb glass	1.54	0.90	126	22.8

Clauer *et al.* (1981) conducted a number of experiments with a Q-switched Nd:glass laser to investigate various influential factors, such as thermal protective coatings, confined overlays and laser power densities, which significantly affect the peak pressure of the pressure pulse for LSP. It was observed that the peak pressure was significantly increased when selecting the confined ablation mode, with the acoustic impedance of confined overlay being a key influential factor in the magnitude of the pressure pulse.

Masse and Barreau (1995a) investigated residual stresses in a hypoeutectoid steel (0.55% C) impacted by a pulse pressure of 25 kbar (laser power density of 4 GW/cm^2) with a WCM. It was observed that the surface compressive residual stress was up to 350 MPa in the WCM. Furthermore, if using a glass-confined mode, the laser power density could be reduced to 1.7 GW/cm^2 to achieve the same level of surface compressive residual stress. Above these laser power densities, surface compressive residual stresses were saturated, while the plastically affected depths were in the range 0.9–1.1 mm, depicted in Fig. 2.5.

In addition, it was observed that the mechanical effects of a laser-induced stress wave in a metal alloy depend significantly on whether the material is covered by a thermal protective material or absorbent coating (Peyre and Fabbro, 1995b; Fabbro *et al.*, 1998). In the direct ablation mode, the heated zone caused by the thermal effect is compressively plasticised by the surrounding material during the dilatation. As a result, tensile strain and stresses may occur after cooling. If the metal surface is coated with a thermal protective material (black paint or Al foil), the thermal effect only occurs in the coating layer. Shock waves penetrate into the material to create a pure mechanical effect. After the laser pulse duration, the surrounding material reacts to the volume expansion of the treating zone, inducing a compressive stress field (Peyre and Fabbro, 1995b).

The thermal protective materials may be metallic foils (aluminium foil) or organic paints (black paint) or adhesives. Coating on a metal surface not only protects the surface from radiation but also enhances the induced

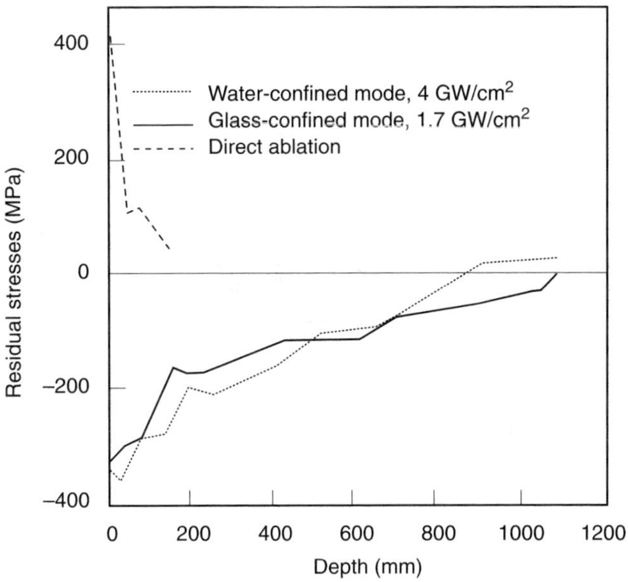

2.5 In-depth residual stress profiles with various treatments (Masse and Barreau, 1995a).

plasma (Hong *et al.*, 1998; Peyre *et al.*, 1998a, b, c; Clauer and Lahrman, 2001; Auroux *et al.*, 2001). It was observed that the coating could play a fundamental role in plasma properties and the plasma pressure (Peyre *et al.*, 1998a, b, c). In order to increase the magnitude of dynamic stresses in the metal material, the coating layer could combine its constraining characteristics with some impedance mismatch effects (Peyre *et al.*, 1998b). When using a thick enough coating with low acoustic impedance, a much higher magnitude of dynamic stress than the plasma pressure, compared with that in the uncoated material, could be achieved inside the material. For example, when a 316L stainless steel was covered with a 100 μm thick Al-based coating, the pressure magnitudes for the bare 316L stainless steels were very similar to those measured on the Al targets, but with the coating, the peak stress levels inside the steels were increased by more than 50%, shown in Fig. 2.6. These impedance mismatch effects would allow the use of lower laser power densities.

However, such thermal protective coatings must have very good adhesive properties, especially for multiple shock loadings (Peyre *et al.*, 1998a; Fabbro *et al.*, 1998). Peyre *et al.* (1998a) concluded three main roles of protective coating (100 ± 30 μm thick Al adhesives) on the 316L stainless steels. Firstly, the coating can protect the component to avoid ablation from thermal effects. Secondly, the amplitude of stress waves can be increased by up to 30–50%. Thirdly, the resultant stresses were scattered owing to interface mismatch effects between steel substrate and Al adhesives.

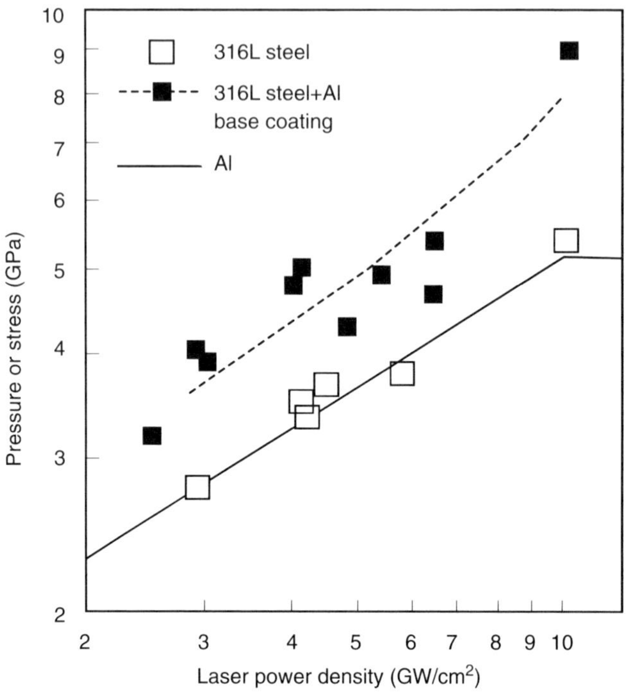

2.6 Peak pressure levels induced by LSP in 316L stainless steels with or without a 100 µm Al foil coating (Fabbro *et al.*, 1998).

Hong *et al.* (1998) discovered that black paints and Al foils differ considerably in their absorption ability when a high-intensity laser irradiates the material surface. In their experimental conditions, the black paint layer could absorb almost all the laser energy, while the Al foil layer could only absorb 80%. They also pointed out that the magnitude of the pressures was significantly increased with the thermal protective overlay, compared to the bare material surface.

Fabbro *et al.* (1998) investigated the effects of the impedance mismatch of the coating on Al components in LSP. In their experiments, 250 µm thick Al components were covered with 5 µm coatings of four different metallic materials (Al, Ta, Mo and Cu) and an Al-based paint. The components were then irradiated by laser at two power densities, 1 GW/cm^2 and 4 GW/cm^2, respectively. The coatings were assumed to be thick enough to avoid their complete ablation and to form the plasma completely, but thin enough to minimise impedance mismatch effects. Despite a slight difference noticed in the material with the Cu coating, shown in Fig. 2.7, probably resulting from a smaller absorbed intensity, the results reveal that the laser-induced pressure was quite independent of the nature of the coating material.

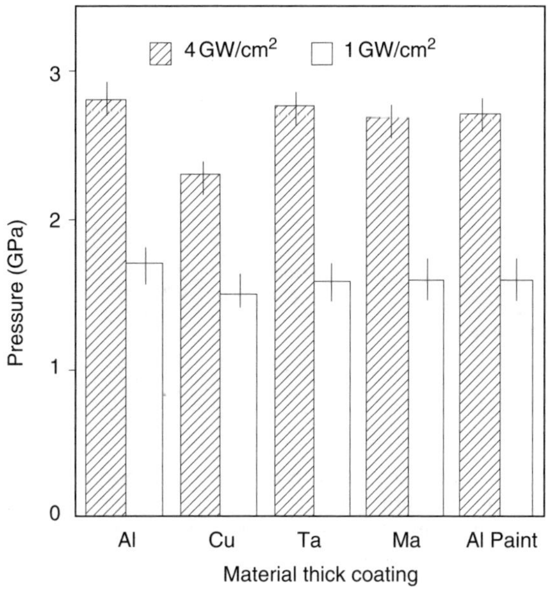

2.7 Peak pressures obtained from different coating materials with laser power densities of 1 and 4 GW/cm^2 (Fabbro *et al.*, 1998).

The thermal protective coating is used not only to protect the substrate from the thermal effects of ablation but also to increase the amplitude of the stress waves (Clauer *et al.*, 1981; Peyre *et al.*, 1998a, b, c). Peyre *et al.* (1998b) investigated the distribution of surface residual stresses in notched fatigue samples (55C1 steel) with or without coating. The results in Fig. 2.8 show that the uncoated material has high tensile residual stresses even when confined with water. These tensile stresses were attributed to severe surface melting, which confirms that the overall role of coatings is to preserve the surface integrity. In contrast, high compressive residual stresses on the surface were achieved when the materials were coated with the aluminium paint.

2.5.3 Laser spot size and laser duration

The laser spot diameter can be varied and is limited only by the power density and laser power required. Varying the spot diameter from 1.2 mm to 5 mm affected the propagation behaviour of shock waves in 55C1 steel foil specimens 620 μm thick (Fabbro *et al.*, 1998). For a small diameter, the shock wave expanded like a sphere, which resulted in attenuation at a rate of $1/r^2$, while for a large diameter, the shock wave behaved like a planar front, which attenuated at a rate of $1/r$. The net result was that the energy

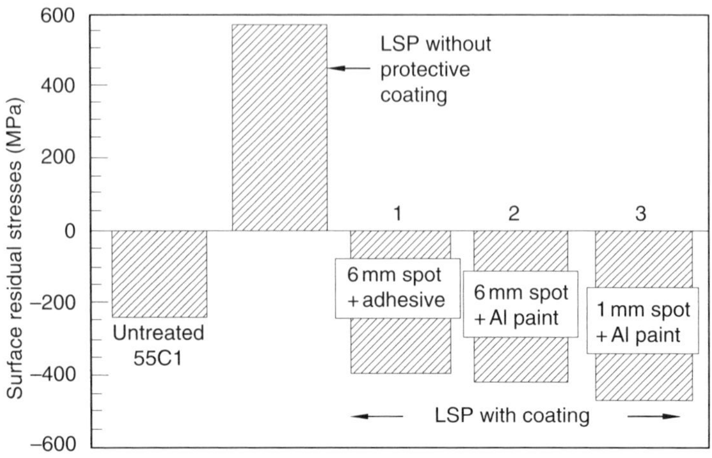

2.8 Surface residual stresses with different LSP conditions at 5 GW/cm² in a water-confining mode, (1) 6 mm impact + aluminium adhesive, (2) 6 mm impact + aluminium paint, and (3) 1 mm impact + aluminium paint (Peyre *et al.*, 1998b).

attenuation rate was less for the large diameter, and the planar shock wave can propagate further into the material. This was also seen in shock wave propagation in rock where the shock wave from a 10 J pulse was thought to decay like a spherical shock wave over 10 mm (Bolger *et al.*, 1999). The shock wave from the 100 J pulse behaved like a planar shock wave and propagated 25 mm into the rock.

Peyre *et al.* (1998b) investigated residual stresses in a 55C1 steel with respect to changes in laser spot sizes, and the laser spot diameters used for LSP were 1 mm and 6 mm, respectively. The results indicate that the large spot size produces a residual stress much deeper below the treated surface than the small one. However, the magnitude of surface compressive residual stresses was not increased as a result of the large spot size.

The magnitude of residual stresses is usually high at the surface and gradually decreases with depth (Fairand and Clauer, 1978; Clauer *et al.*, 1992). However, when using a circular laser spot, residual stresses at the centre of spot are unstable owing to the complicated interaction of shock waves in this region (Masse and Barreau, 1995a). Such a phenomenon can be artificially minimised by changing the geometry of the laser spot, i.e. a square or an ovoid one (Ballard *et al.*, 1991; Masse and Barreau, 1995a; Peyre *et al.*, 1996; Clauer, 1996; Clauer and Lahrman, 2001).

The diameter of the laser spot in practice usually ranges between a few hundred micrometres and 6–10 mm. For example, when using various diameters of laser spot to treat a 55C1 steel under the same LSP conditions, surface residual stresses were nearly the same, but the plastically affected depth with compressive residual stress tended to decrease drastically with a small impact size (<1 mm) (Peyre *et al.*, 1996).

A laser system normally delivers two kinds of temporal shapes of laser pulse, a Gaussian pulse shape and a short rise time (SRT) pulse shape. Devaux *et al.* (1993) first conducted a series of experiments to investigate the effects of breakdown phenomenon caused by two temporal shapes of laser pulse. They noticed that the use of a SRT laser pulse for LSP could reduce the effects of breakdown and obtain a much higher pressure on the metal surface. They also observed that the breakdown threshold of the laser power density was strongly dependent on the rise time of the laser pulse in different confined modes. In their experimental measurements, the breakdown occurred at 3 GW/cm^2 for a 30 ns Gaussian pulse with a WCM, but at 8 GW/cm^2 for a 3 ns Gaussian pulse with a glass-confined mode (GCM).

As a laser system can deliver a wide range of pulse duration (between 0.1 and 50 ns) for LSP, the laser pulse duration directly controls the pressure pulse duration (Devaux *et al.*, 1993; Gerland and Hallouin, 1994; Couturier *et al.*, 1996; Cottet and Boustie, 1989). For instance, Fabbro *et al.* (1998) reported that a higher pressure was achievable with a short duration of laser pulse. Typically, a laser duration of 0.6 ns with a laser power density of 100 GW/cm^2 can generate a peak pressure of 10 GPa.

At a constant level of pressure, the shorter pressure duration tended to generate a higher magnitude of residual stresses (Fabbro *et al.*, 1998; Noack *et al.*, 1998; Peyre *et al.*, 2000a). A 12% chromium stainless steel was impacted by a laser pulse with three different periods of pulse duration, 0.6, 2.3 and 20 ns, corresponding to three different laser power densities of 7, 40 and 140 GW/cm^2, respectively (Fabbro *et al.*, 1998). It was observed that the maximum surface residual stress was achieved with a pulse duration of 2.3 ns and a laser power density of 40 GW/cm^2.

2.5.4 Laser power density and wavelength

The magnitude of surface residual stresses increases with the magnitude of the plasma pressure, which is related to the incident laser power density. When the laser power density exceeds a threshold, residual stresses increase with depth but decrease at the surface because of surface release waves. This indicates that there are optimal LSP conditions. For instance, surface compressive residual stresses in an A356-T6 alloy specimen increased up to 145 MPa for an increase in laser-induced pressure from 1.3 to 1.5 GPa when the laser power density was changed from 1.5 to 2 GW/cm^2. However, a further power density increase to 3 GW/cm^2 tended to reduce the stress level to 100 MPa, whereas the in-depth compressive residual stress continued to increase (Peyre *et al.*, 1996).

Peyre *et al.* (1996) later found that the laser-induced pressure was a function of the laser power density of the laser pulse with two different temporal shapes for WCM. They reported that the saturated pressure was reached at a laser power density of 4 GW/cm^2 when using a Gaussian pulse for LSP. The saturated pressure from the Gaussian pulse was explained in terms of a reflecting dielectric breakdown phenomenon in the confinement medium that limits the amount of energy reaching the metallic surface (Peyre *et al.*, 1996; Fabbro *et al.*, 1990; Berthe *et al.*, 1997; Sollier *et al.*, 2001). However, the breakdown threshold of the laser power density was increased up to 10 GW/cm^2 when using a SRT laser pulse for LSP (Peyre *et al.*, 1996).

Berthe *et al.* (1999) first conducted studies on characterisation of laser shock waves and breakdown phenomenon with respect to changes in laser wavelength from IR to UV. The results indicated that the pressure produced by the laser pulse with wavelengths of 0.532 μm (green) and 0.355 μm (UV) had a similar profile to that generated with a wavelength of 1.06 μm (IR). In addition, the pressure produced by a laser pulse with a wavelength in the IR, corresponding to a laser power density of 10 GW/cm^2, was saturated at 5.0 GPa with WCM. However, the saturated pressure at UV and green wavelengths occurred at higher laser power densities than that at IR wavelengths. Moreover, the pressure duration at UV wavelengths decreased more significantly than that at IR wavelengths when the laser power density is increased.

2.5.5 Multiple laser shock peening

Compressive residual stress can be driven deeper below the surface by using successive shocks. Clauer (1996) conducted multiple LSP on a 0.55% carbon steel. It was observed that the plastically affected depth with compressive residual stress increased linearly with the number of impacts on the same spot, described in Fig. 2.9. When the number of shots was increased from one to three, the plastically affected depth (L_p) was increased from 0.9 to 1.8 mm.

Peyre *et al.* (1996) also studied residual stress profiles in three aluminium alloys (7075, cast A356 and Al12Si) induced by multiple LSP. The results show that, on the cast A356 alloy, no increases in surface compressive residual stress can be achieved with two or three impacts at the same spot, but the plastically affected depths are significantly increased. However, for the 7075-T7351 aluminium alloy, multiple LSP has a clear beneficial effect on the magnitude of surface residual stresses (–150 MPa for one impact and –300 MPa for three impacts), indicated in Fig. 2.10.

Figure 2.11 shows results of two successive LSP at the same spot on a Ti-6Al-4V titanium alloy. It can be observed that two successive laser shots generate residual stresses deeper than a single shot (Dane *et al.*, 1997). The

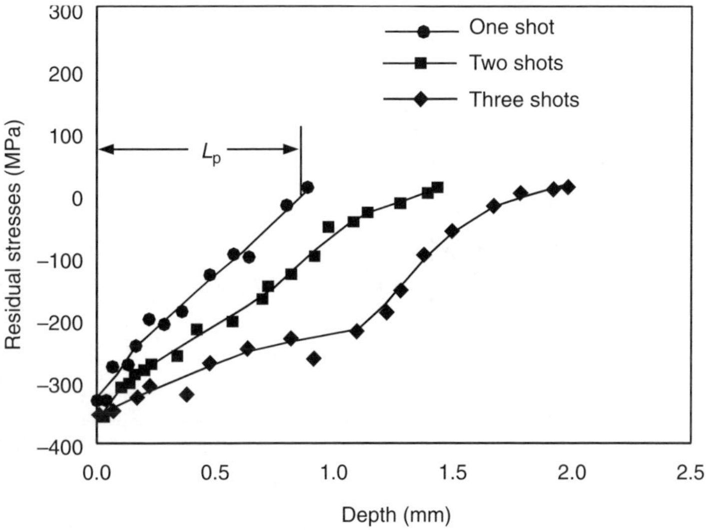

2.9 In-depth residual stress profiles induced by multiple impacts on a 0.55% carbon steel (Clauer, 1996).

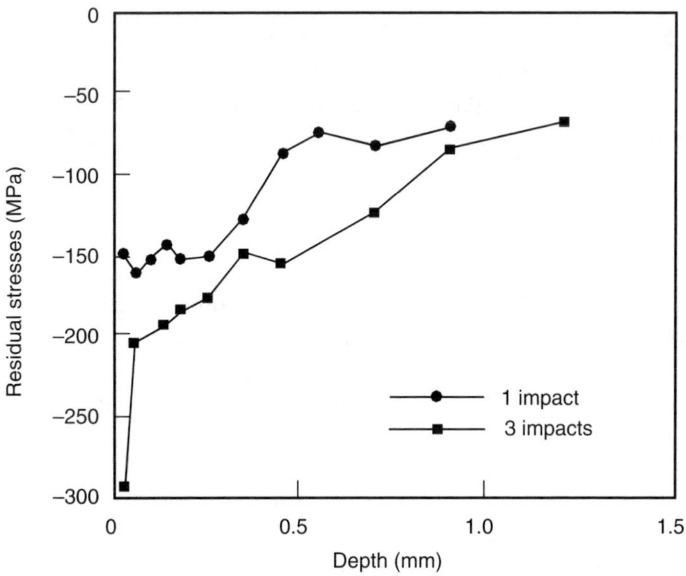

2.10 In-depth residual stress profiles induced by multiple LSP on 7075-T7351 Al alloy (Peyre *et al.*, 1996).

same trend was observed in other alloy systems. For a 0.55% carbon steel, as the number of shots on the surface were increased from one to three, the depth of compressive residual stresses was increased from 0.9 to 1.8 mm (Masse and Barreau, 1995a). For a 7075 aluminium alloy, multiple LSP had

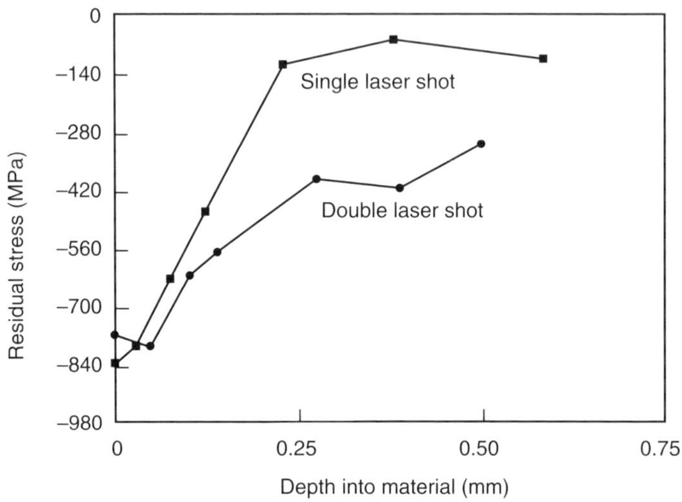

2.11 Comparison of residual compressive stresses induced by two successive laser shots and one single laser shot in Ti-6Al-4V titanium alloy using a laser pulse energy density of 200 J/cm^2 and a pulse duration of 30 ns (Dane *et al.*, 1997).

a clear beneficial effect on compressive residual stress levels on the surface. A treatment of 4 GW/cm^2 in laser power density generates a residual compressive stress of 170 MPa after the first impact, 240 MPa after the second impact and 340 MPa after the third impact (Peyre *et al.*, 1996).

2.5.6 Overlapping of laser spots

Owing to power density requirements and laser power availability, even laser systems with the potential for 200 J per pulse will be limited by the area covered by the laser spot. Overlapping laser spots is the method used to treat large areas in practice. In the traditional SP process, the coverage ratio in the processing area is known to be a very important factor for optimising the residual stress field in the material (Chu *et al.*, 1995). The coverage ratio is normally referred to as the ratio between the overlapping area and the impact spot size for two successive SP or LSP processes (Prevéy and Cammett, 2002). For LSP, many studies (Peyre *et al.*, 1996; Masse and Barreau, 1995a; Fabbro *et al.*, 1998; Clauer, 1996) show that an increase in the coverage ratio increases the plastically affected depth. Optimisation of coverage ratio could lead to a better treatment result for many metallic materials. For example, in order to minimise disadvantage associated with a circular laser spot, multiple impacts with a coverage ratio of 50–70% to overlap circular spots on the surface are commonly utilised (Peyre and Fabbro, 1995b; Fabbro *et al.*, 1998; Masse and Barreau, 1995a).

Overlapping of laser spots has been investigated for the treatment of larger areas of many industrial metals such as 1026 steel and ductile cast iron (Cai and Zhang, 1996), A356, Al12Si and 7075 aluminium alloys (Peyre *et al.*, 1996), 55C1 steel (Ruschau *et al.*, 1999) and titanium (Zhang *et al.*, 1997). The results show that there was a relatively uniform distribution of compressive residual stresses across overlapped regions after LSP, with no indication of tensile residual stresses in the overlap regions. It indicates that little or no degradation in properties would occur in such processes.

Two laser spot diameters for LSP at the same power density of 5 GW/cm^2 were used to investigate effects of both overlapping and spot sizes on material properties of 55C1 steel (Ruschau *et al.*, 1999). LSP was performed with four impacts of 6 mm in diameter with 50% overlap, and with 50 impacts of 1 mm in diameter with 25% overlap. Residual stress measurements showed that, with a small laser spot, the plastically affected depth was significantly reduced. This is consistent with the results of reduced penetration distance of residual stress for small laser spots (Clauer and Koucky, 1991) when applying LSP to metallic components.

2.5.7 Laser shock peening for thin sections

LSP for thick or thin metal components requires optimisation of the laser pulse that impacts on the metal surface so that the best treatment results can be reached. For a thick component, laser pulses can be used individually or simultaneously to impact on any location on the surface. However, for a thin section, a split laser pulse is required to impact on opposite sides of the thin section to balance the forces generated (Clauer, 1996). If the section by one-sided peening is thin enough, the peened spot can create a dimple on the irradiated side and a bulge on the opposite side. It can also cause spalling and fracture if the shock waves are strong enough. Meanwhile, if the laser pulse impacts on a large area, significant curvatures or other distortions can be induced in the component (Clauer and Lahrman, 2001).

Clauer (1996) conducted an experiment for laser peening a 4340 steel sheet of 1.5 mm in thickness. The sheet was peened with one to five shots from both sides simultaneously. It was observed that the depth of compressive residual stress was about 0.5 mm for one shot and the magnitude of compressive residual stress was higher after five shots. In addition, the tensile residual stress at the mid-plane (0.75 mm) of the sheet was higher after five shots (Fig. 2.12). Clauer and Lahrman (2001) also investigated the compressive residual stress in a 1.0 mm thick section of Ti-6Al-4V. The experiments were conducted with three processing conditions, low (one shot with laser power density 5.5 GW/cm^2), intermediate (three shots with laser power density 5.5 GW/cm^2) and high (three shots with laser power density 10 GW/cm^2). It was observed that the residual stress level was increased by

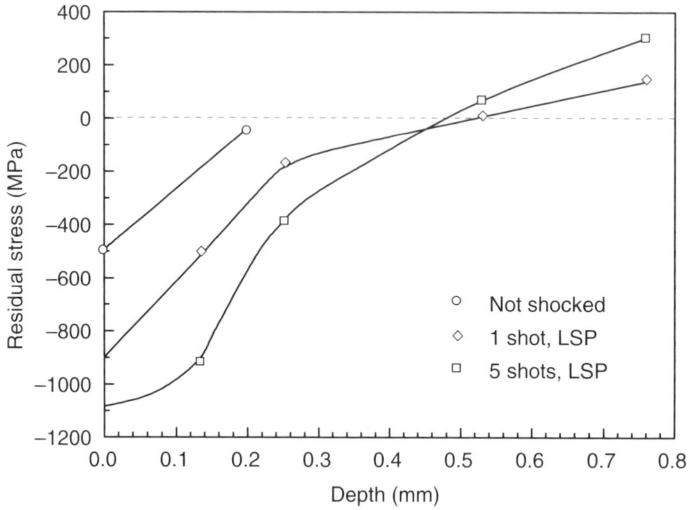

2.12 Residual stress profiles in 4340 steel sheet of 1.5 mm in thickness (Clauer, 1996).

25–30% from one shot to three shots at the same laser power density of 5.5 GW/cm^2. However, if the laser power density was increased from 5.5 to 10 GW/cm^2 as well, the residual stress level was increased by up to 40–50%.

2.5.8 Comparison between laser shock peening and shot peening

In military aircraft and spacecraft, there are a large number of thin metal components, which can be treated using the two-sided peening. For thin sections, the use of SP is not practical because of the potential damage from the process. LSP is considered more suitable for thin sections (Clauer *et al.*, 1983). The US Air Force conducted a review of both the SP and LSP technologies (Thompson *et al.*, 1997) for practical applications. It focused on the leading edge of a turbine fan blade with an emphasis on reducing high cycle fatigue failure caused by foreign object damage. SP was seen to have several limitations for high cycle fatigue. Shot peened blades did not meet the fatigue lifetime requirements. In particular, SP generates a rough surface with large increases in the value of the mean and peak roughness. This can be advantageous for paint adhesion but is detrimental to wear and fatigue properties. The roughening effects of LSP and SP on A356 and 7075 aluminum alloys are shown in Table 2.3. However, for wear applications, removal of the roughened surface is a necessity; owing to the limit thickness of the shot peened compressive layer, removal of the rough surface also leads to significant reduction in the surface layer thickness of compressive residual stress.

Table 2.3 Comparative roughening effects of LSP and SP (Peyre *et al.*, 1996)

Material and processing	R_a (μm)	R_t (μm)
A356 as milled	0.7	6.2
A356 LSP (2 GW/cm^2, 2 impacts)	1.1	7.5
A356 shot peening (F38-50N, 0.3 mm beads)	5.8	33
7075 as milled	0.6	5.2
7075 LSP (4 GW/cm^2, 3 impacts)	1.3	11
7075 shot peening (20–23A, 125%, 0.6 mm beads)	5.7	42

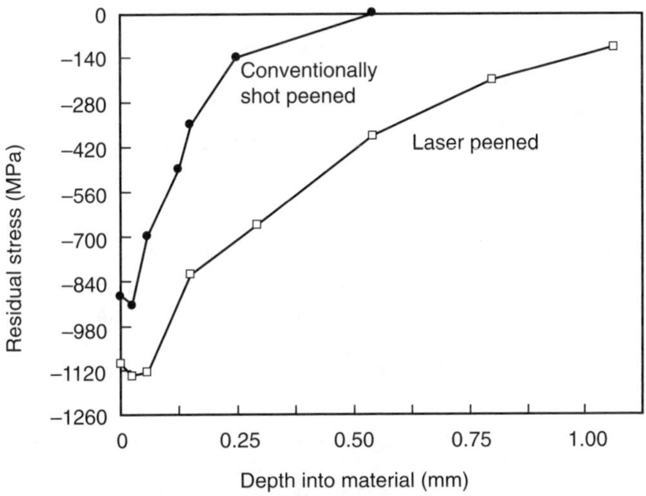

2.13 Residual stresses in the surface of Inconel 718 induced by laser peening and conventional shot peening (Dane *et al.*, 1997).

The actual depth of LSP-induced stresses is dependent on processing conditions and material properties and generally ranges from 0.5 mm to over 1 mm. Moreover, small surface stress gradients are observed after LSP, which is beneficial because it is known to be important in reducing or eliminating cyclic stress relaxation. Residual stresses induced by LSP in Inconel 718 alloy are compared with the typical results achieved by conventional SP in Fig. 2.13 (Dane *et al.*, 1997). Clearly, residual stresses are much deeper for LSP than for conventional SP.

The quantitative comparison between the loading conditions induced by LSP with a water overlay and those induced by SP is presented in Table 2.4. The most distinctive change in the impact conditions involves the duration of the induced peak pressure, which is 10–20 times longer in the case of SP. Shot-peened surfaces are subjected to more multiaxial, intense loadings than the laser-peened surfaces. An integrated approach combining LSP with

Table 2.4 Comparative loading conditions induced by LSP and SP (Peyre *et al.*, 1996)

Process	Peak pressure (GPa)	Diameter of impacts (mm)	Pressure duration (μs)	Mechanical impulse (GPa μs)	Strain rate (s^{-1})
LSP	0–6	1–15	0.05	0–0.3	10^6
SP	3–10	0.2–1	0.5–1	1–10	10^4

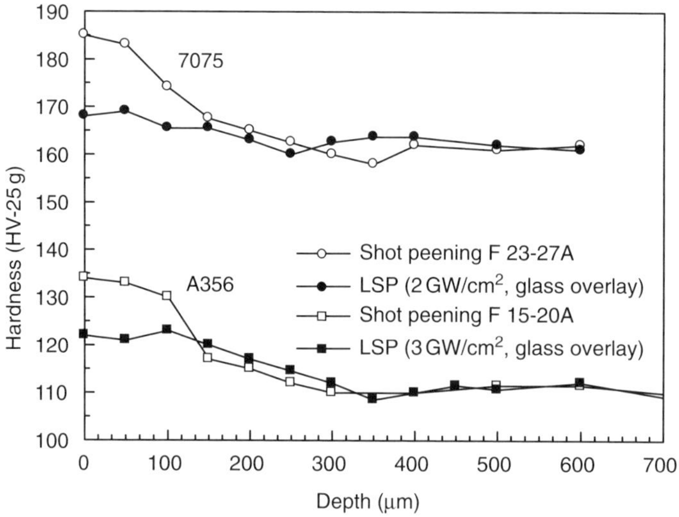

2.14 Vickers hardness measurements with a 25 g load on A356-T6 and 7075-T7351 aluminium alloys treated by laser peening and shot peening respectively (Peyre *et al.*, 1996).

SP for a 7075 aluminium alloy specimen showed that such a combination could result in enhanced properties with increases in both in-depth and surface compressive residual stresses (Peyre *et al.*, 1996).

For both LSP and SP, the shock hardening effect below the surface decreases with increasing distance from the surface. Peyre *et al.* (1996) compared the effects of LSP and SP on the surface hardness for 7075 and A356 aluminium alloys, shown in Fig. 2.14. SP resulted in twice the surface hardness increase compared with LSP. This was attributed to the longer application of pressure in SP, which promotes greater dislocation generation and motion. The number of slip planes activated by multiaxial surface loading in SP may affect the hardness. However, some fundamentals concerning these mechanisms need to be investigated further.

2.6 Modifications in surface morphology and microstructure

The surface morphology of metals has a great effect on fatigue behaviour. Many investigations related to the surface morphology of laser-peened materials have been performed with scanning electron microscopy (SEM) observations and roughness measurements. When no protective laser-absorbent coating was used on the material surface, LSP can cause severe surface melting and vaporisation, particularly in aluminium (Clauer *et al.*, 1976; Gerland and Hallouin, 1994). This can result in resolidified droplets and craters leading to very rough surfaces. These problems can be solved with energy absorbent coatings as discussed in previous sections. However, there has been no systematic fundamental understanding in the LSP literature of the interaction of the microstructure with laser-induced shock waves and the resulting changes in the microstructure.

The LSP process is not a thermal process but a mechanical process for metallic materials and it is accompanied by significant changes in microstructures and phases. These changes have been investigated by means of transmission electron microscopy (TEM), SEM and X-ray diffraction analysis. Microstructural changes induced by LSP have been related to the laser parameters and the treatment conditions of the alloys. In laser peened aluminium alloys such as welded 5086-H32, 6061-T6 (Clauer *et al.*, 1976) and 2024-T62 (Zhang and Yu, 1998), it was observed that the dislocation density increased significantly. High dislocation densities were also a prominent microstructural feature in low carbon steels after LSP (Peyre *et al.*, 1998b; Atshulin *et al.*, 1990). LSP of Hadfield manganese steel was found to induce extensive formation of ε-hexagonal close-packed (ε-hcp) martensite and high density dislocations in the γ-face-centred cubic (γ-fcc) austenite matrix (Chu *et al.*, 1995).

Investigations of the effect of LSP on weld zones in 18Ni (250) maraging steel showed that, after the LSP treatment, the austenite weld phase reverted to martensite and the dislocation density qualitatively increased in the martensite matrix (Banas *et al.*, 1990a, b). Numerous twins as well as α-phase embryos located at the twin intersections were found in laser peened 304 austenitic stainless steel (Devaux *et al.*, 1993) and in 316L stainless steel (Gerland and Hallouin, 1994). A Fe–Ni alloy was processed using LSP with a laser power density of 100 GW/cm^2 and 10 TW/cm^2 without transparent overlay and absorbent coating (Grevey *et al.*, 1992). Very thin twinned grains were found on the surface because of melting and rapid solidifying. A martensite transformation zone was observed at the back face of laser peened Fe–Ni alloy sample caused by reflection of shock wave from the back face.

The minimal change in hardness of the bulk material outside the heat affected zone (HAZ) from a laser peen pulse of 3.5 GPa in peak pressure has been noted before (Clauer and Fairand, 1979). LSP has been reported to improve the hardness of underaged 2024-T351 but not peak aged materials like a 2024-T851, 7075-T651 or 7075-T73. However, for the 6061-T6 specimens, there was no change in hardness or strength reported (Fairand *et al.*, 1976). It was hypothesised that the precipitation hardening in the T6 condition is significantly large enough to mask any shock wave strain hardening. It was also hypothesised that exceeding a threshold shock wave pressure of 7.5 GPa was required to change the properties of peak aged aluminium alloys significantly (Fairand and Clauer, 1978). Nevertheless, from the investigation of property changes with micro/nano-indentation (Montross *et al.*, 2000), a shock wave with a pressure of 6 GPa was sufficient to increase the hardness in the bulk 6061-T6 material significantly.

2.7 Effects on mechanical properties

Many materials display pronounced improvements in fatigue life with LSP. The beneficial effects of LSP may originate from surface compressive stresses in the large affected depth and improved surface quality, which delay the development of fatigue cracking. Investigations of several different aspects of the fatigue behaviours, such as fatigue life, fatigue strength and fretting fatigue, have been reported.

2.7.1 Fatigue life and strength

Investigations of aluminium alloys, steels and titanium alloys have shown that LSP can increase the fatigue strength of these materials. The initial research work focused on the effect of LSP on the fatigue crack growth of pre-existing cracks, using different laser spot shapes on the pre-cracks (Clauer *et al.*, 1983). The two different laser spot configurations were applied around a hole in a 2024-T3 aluminium alloy specimen, shown in Fig. 2.15(a). The specimen had a cantered hole with small starter notches machined into its sides. The region around the hole was shocked simultaneously on both sides using split laser beams. One laser spot configuration used a solid spot to treat the entire region around the hole, while the other laser spot used only an annular-shaped area around the hole and notched region. The fatigue life increases for both laser spot configurations, shown in Fig. 2.15(b), and the laser-peened 2024-T3 specimens with the solid laser spot had a fatigue life about 40 times longer than the non-shocked ones, whereas those with the annular laser spot had a life about three times longer than the non-shocked ones.

Further studies confirmed the beneficial effect of LSP on fatigue cracking resistance of pre-existing cracks that were effectively arrested by LSP

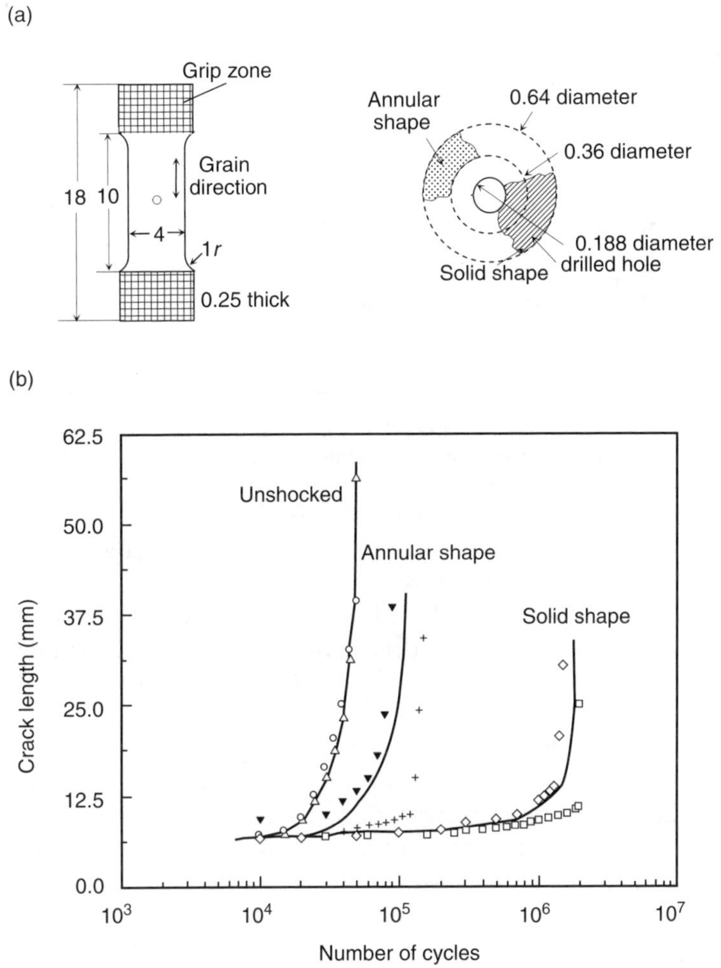

2.15 Increased fatigue life in 2024-T3 aluminium after laser peening (Clauer *et al.*, 1983). (a) Specimen configuration and laser shocked region shape (dimensions in inch). (b) Fatigue life. Laser energy densities for two sides of laser peened specimens are: 75 and 75 J/cm^2 (▼); 81 and 80 J/cm^2 (+); 82 and 78 J/cm^2 (◇); 84 and 78 J/cm^2 (□).

(Vaccari, 1992; Clauer, 1996; Yang *et al.*, 2001). The laser peened specimens with pre-crack have fatigue lives in almost the same range as those of the laser peened materials without a pre-crack (Cai and Zhang, 1996; Clauer, 1996).

Fatigue tests on A356-T6, Al12Si-T6 and 7075-T7351 Al alloys after LSP treatments reveal that improvements in fatigue strength are approximately +36% for A356-T6, +22% for both Al12Si-T6 and 7075-T7351 Al alloys, after a fatigue life of up to 10^7 cycles (Peyre *et al.*, 1996). Most of the Al alloys, such as steels and titanium alloys treated by LSP, were shown to have

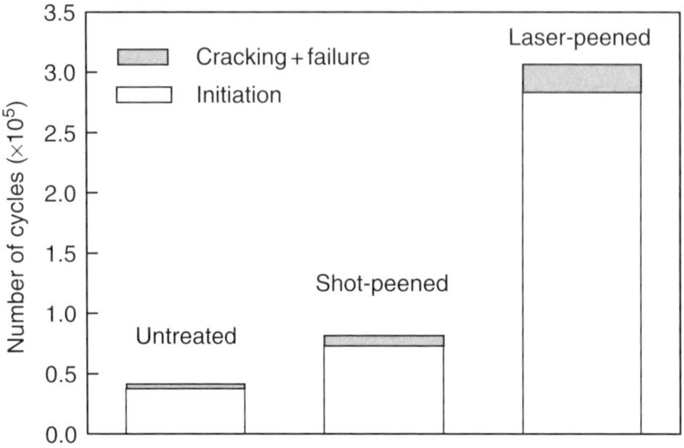

2.16 Comparison of crack initiation and crack growth stages at σ_{max} = 260 MPa for crack detection tests on 7075-T7351 Aluminium alloy (Peyre *et al.*, 1996).

better fatigue performance than those treated by SP (Ballard *et al.*, 1991; Clauer, 1996; Peyre *et al.*, 1998a, b; Zhang *et al.*, 1999). This improvement under LSP treatment is attributed to the higher level of compressive residual stress and the greater plastically affected depth in the materials as well as the preservation of surface roughness of the materials (Fabbro *et al.*, 1998).

Peyre *et al.* (1996) also examined and compared effects of LSP and SP on early and later stages of crack propagation for a 7075-T7351 aluminium alloy. The notching process was used to localise any crack initiation to the notch root, with a stress concentration of K_t = 1.6–1.7. The LSP consisted of three square laser spots with 50% overlap for the cast alloys but about 67% for the 7075 alloy specimens. Fatigue testing was done under three-point bending with a stress ratio, R = 0.1 at 40–50 Hz. It can be seen from Fig. 2.16 that, for an applied stress of 260 MPa, the laser peened specimens dramatically improved the fatigue life. There were clear differences between specimens treated by LSP and SP in the early and late stages of crack growth. Compared to the as-received specimens, the fatigue life improvements from LSP can be separated into a seven-fold increase in the early crack growth stage and only a three-fold increase in the later propagation stage. In contrast, SP only provided a homogeneous two- to three-fold increase in both the early and late stages of crack growth when compared to the as-received specimens. The difference in results between LSP and SP was attributed to surface embrittlement and surface roughening due to the SP process that creates sites where cracks develop rapidly and tends to reduce the beneficial effects of compressive residual stress. The bending fatigue properties (Peyre *et al.*, 1996) of a 7075-T7351 aluminium alloy that has received LSP and SP, respectively, are compared in Fig. 2.17.

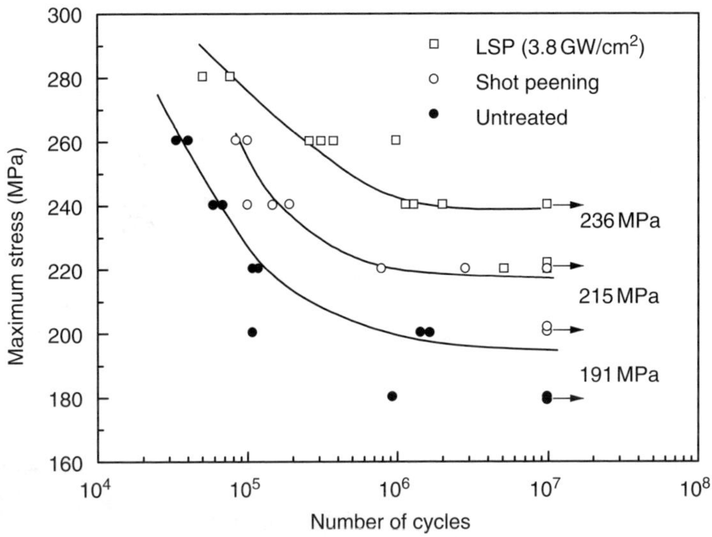

2.17 *S–N* curves for untreated, shot peened and laser peened 7075-T7351 alloys (Peyre *et al.*, 1996).

SP provided only an 11% increase in the fatigue strength at 10^7 cycles, while LSP provided a 22% increase, compared with the as-received untreated specimens. This improvement was explained by the greater depth of the residual compressive stress field induced by LSP compared with SP.

Peyre *et al.* (1998b) further investigated the effect of different laser spot sizes and configurations on the fatigue behaviour of steel and aluminium alloys. Notched specimens were again used as described previously but fatigue testing was conducted by four-point bending of notched specimens with $R = 0.1$. For the 55C1 steel specimens, the small 1 mm laser spot had a greater reduced residual stress depth compared with the large 6 mm spot. However, the surface residual stresses were approximately the same for both 1 and 6 mm spots with the same coating and processing parameters. The fatigue specimens were subjected to LSP with either four, 6 mm diameter laser spots with 50% overlap or fifty, 1 mm diameter spots with 25% overlap. From Fig. 2.18, it can be seen that the overlapped small diameter (1 mm) laser spots displayed an approximately equivalent improvement in fatigue strength (490 MPa) at 2×10^6 cycles compared with the overlapped large ones (6 mm) of 470 MPa. This is a significant improvement from the fatigue strength of 380 MPa at 2×10^6 cycles for the as-received material. This also indicates that LSP with small impacts could be considered a potential method for improving the fatigue life of structural components without having to use larger laser systems that are more costly and more difficult to control.

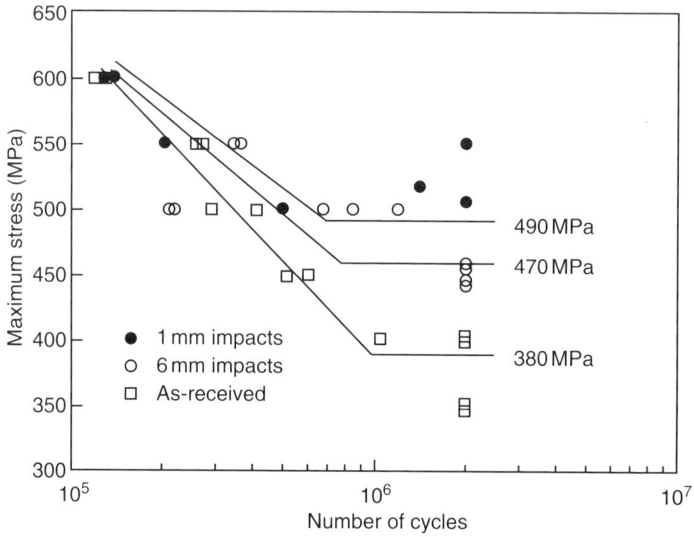

2.18 *S–N* curves of notched bending 55C1 steel samples treated by laser peening (Ruschau *et al.*, 1999).

In an industrial magazine, LSP was reported to have improved the fatigue life of notched, 1.5 mm thick AISI 4340 hardened steel specimens that had a Rockwell hardness of $R_c = 54$ before LSP (Scherpereel *et al.*, 1997). With 7.62 mm for both notch radius and notch thickness, fatigue specimens after LSP with a laser spot size of 9.91 mm showed that the fatigue strength was increased by 60–80%, from approximately 552–612 MPa to approximately 966–1035 MPa, although other details such as stress ratio and testing frequency were not given.

LSP also successfully improved the fatigue performance of titanium alloys, such as Ti-6Al-4V used in turbine compressors and Inconel superalloys used in turbine hot sections. Initial tests of laser peened blades showed a 10–40% improvement in fatigue strength, allowing engines to operate at higher loads (Dane *et al.*, 1997). LSP also significantly increased the resistance of titanium fan blades to early fatigue failure caused by foreign object damage (FOD). After LSP, the fatigue life of damaged blades was found to be equal to or even higher than that of undamaged blades without LSP (Obata *et al.*, 1999).

LSP can also increase the fatigue strength of welds. The investigation of fatigue strength of welded joints of 18Ni(250) maraging steel, showed an increase of 17% in fatigue strength at 2×10^6 cycles by laser shock hardening the heat-affected zones (Peyre *et al.*, 1998b). Meanwhile, the fatigue life of 5456 aluminium alloy welds was also extended by LSP (Clauer *et al.*, 1981). For an applied stress magnitude of 158 MPa, the fatigue life was increased from less than 50 000 cycles for untreated specimens to more than 3 to 6 million cycles without failure for laser peened specimens.

2.7.2 Fretting fatigue

The fretting fatigue properties were investigated on laser peened 7075-T6 aluminium alloy (Clauer and Fairand, 1979) using dog-bone-type specimens as shown in Fig. 2.19(a). Both sides of the regions around the fastener hole in the specimen were laser peened simultaneously and the pad was laser peened with 13 mm diameter laser spots. The fretting fatigue results shown in Fig. 2.19(b) indicate that, at a stress magnitude of 96.6 MPa, the fretting fatigue life was increased by at least two orders of magnitude. At the highest stress level of 110.4 MPa, the fatigue life was still twice that of the untreated specimens.

2.7.3 Stress corrosion cracking

The corrosion behaviour of laser peened materials has been addressed by many researchers, such as Clauer *et al.* (1981) for 2024-T351 Al alloy, Sano *et al.* (1997) for AISI 304 steel, Fouquet *et al.* (1991) for AISI 1010 and Peyre *et al.* (2000c) for AISI 316L steel. The investigation of the corrosion behaviour in these laser peened materials has identified that LSP can be a useful tool to improve corrosion and increase resistance to stress corrosion cracking (SCC). For instance, potentiodynamic tests on laser peened specimens of 2024-T351 Al alloy (Clauer *et al.*, 1981) revealed that the anodic current density was shifted and the passive current density was reduced after LSP. Such behaviour indicates an increase in corrosion resistance of the surface.

Scherpereel *et al.* (1997) investigated the SCC resistance of two stainless steels treated by LSP – one austenitic (AISI 316L) and one martensitic (Z12 CNDV 12.02). In a solution of NaCl 0.01 M + Na_2SO_4 0.01 M, open circuit and polarisation techniques were used to determine electrochemical parameters such as free and pitting potentials and passive current densities in the metals. LSP was found to be more effective for the austenitic (316L) than for the martensitic stainless steel. Although the pitting potentials of the steel were not modified by LSP, the free potentials were shifted to anodic values and the passive current densities were reduced.

LSP was investigated as a way to improve the SCC resistance of thermally sensitised (620 °C for 24 h) type 304 stainless steel, 20% cold worked to simulate neutron irradiation damage (Obata *et al.*, 1999). A creviced bent beam specimen was used to induce a tensile strain of 1% on the surface with the specimen subsequently corroded in water at 288 °C within an autoclave for 500 h. LSP was found to be remarkably more effective than con-

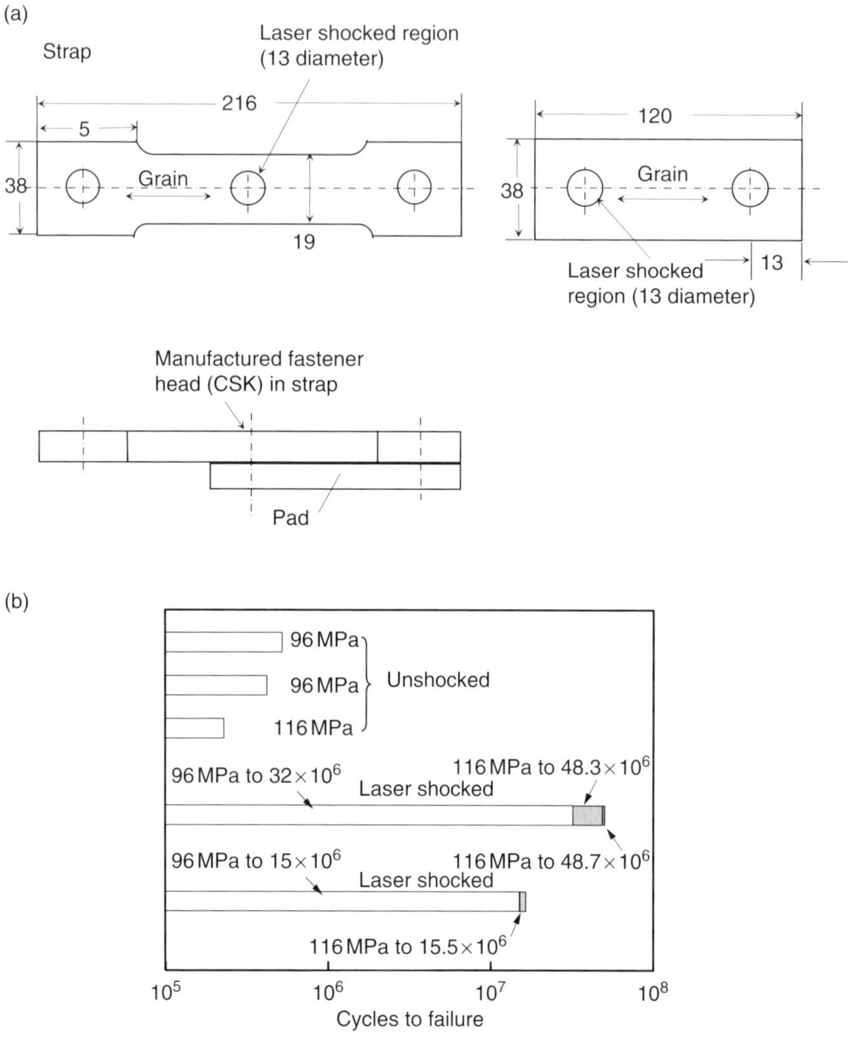

2.19 Increased resistance to fretting fatigue around fastener holes after laser peening of 7075-T6 aluminium (Clauer and Fairand, 1979). (a) Fretting fatigue specimen configuration (dimensions in mm), (b) fretting fatigue results.

ventional SP in increasing the SCC resistance of the thermally sensitised type 304 stainless steel.

2.7.4 Hardening and strengthening

The LSP process can improve the metal surface hardness over the entire region of the laser irradiated area. The magnitude of surface hardening depends on the LSP conditions, alloy type and microstructure of the alloys. It was found that LSP can be used effectively to harden the weld zones of some alloys, such as welded 5086-H32 and 6061-T6 aluminium alloys (Clauer *et al.*, 1976; Ballard *et al.*, 1991) and 18Ni(250) maraging steel (Brown, 1998). LSP increased the yield strength of welded 5086-H32 aluminium alloy to the level of the parent material and increased the tensile yield strength of the welds by 50% for 6061-T6 aluminium alloy (Fairand and Clauer, 1977). The significant strengthening of weld zones in both alloys was noted as due to the higher density dislocations induced by LSP.

In classical shock conditions (ns and GW/cm^2 ranges), surface hardening on peened materials, such as Al-based alloys (Clauer *et al.*, 1981), A356 and 7075 alloys (Peyre *et al.*, 1996), which was mainly ascribed to an increase in dislocation density, is limited to the range +10 to +20% increase in Vickers hardness (HV). The LSP specimen shows a lower HV value than the SP one. This behaviour may be explained in three ways: (1) shock duration is small, so that hardness cannot initiate and propagate inside materials, (2) compared with the SP process, no contact deformation occurs (no Hertzian loading) and (3) impact pressures are usually lower than those from SP (Fabbro *et al.*, 1998).

For LSP on very specific metals, such as Austenitic 304 (Clauer *et al.*, 1981; Gerland *et al.*, 1992) or 316L stainless steel (Peyre *et al.*, 2000b), or under very specific LSP conditions, such as high laser power densities, high pressures (several tenths of GPa) and short pulse durations, large HV increases can be achieved in the materials. Experiments with aluminium alloys showed that the hardness properties of the non-heat treatable (5086-H32) and overaged (2024-T3, 7075-T73) alloys were significantly improved compared with the unshocked properties (Fairand and Clauer, 1978). It was observed that a threshold of laser-induced pressure needed to be exceeded before a clear change in hardness occurred in the treated alloys (Clauer and Fairand, 1979). For an underaged alloy such as 2024-T351, this threshold pressure was approximately 2.5 GPa (Fig. 2.20(a)) while for an overaged 2024-T851, the threshold was 7.5 GPa (Fig. 2.20(b)). Little improvement in the hardness properties of the peak aged aluminium alloys (2024 T8, 7075 T6 and 6061-T6) was noted by some researchers for laser shock pressures of less than approximately 5 GPa. A study of 304 stainless steel also shows that hardness increases with increasing number of multiple shots and further increases are still possible with more shots using a peak shock pressure of 4.9 GPa, as seen in Fig. 2.21. The increase in hardness was reported to be caused by an increase in the dislocation density with increasing laser shock repetitions. However, for thin 2024-T351 aluminium alloy specimens, a hardness peak at mid-thickness was produced by LSP from both sides simultaneously (Clauer and Fairand, 1979). Thus, this split-beam LSP

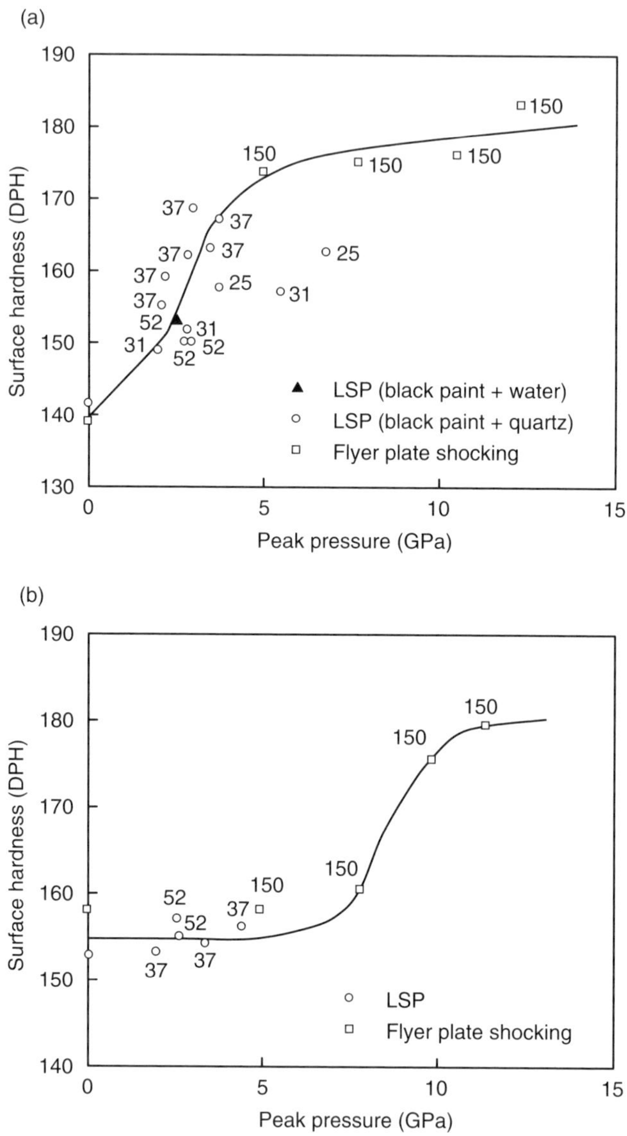

2.20 Dependence of average surface hardness (DPH = diamond pyramid hardness number; the load divided by the surface area of the indentation) on peak shock pressures, comparing laser shocking and flyer plate shocking. The pulse durations in ns are shown beside each data point (Clauer and Fairand, 1979). (a) 2024-T351, (b) 2024-T851.

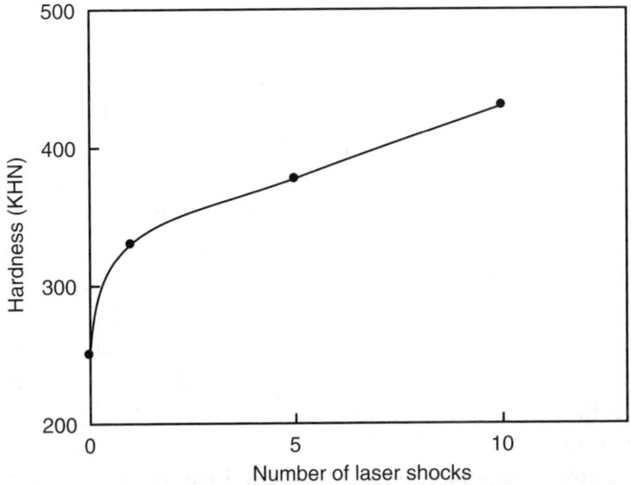

2.21 Increase of surface hardness (KHN = Knoop hardness number; applied load divided by uncovered projected area of indentation) for 304 stainless steel with increasing number of laser shots (Clauer *et al.*, 1981).

procedure produced a more uniform through-thickness hardening in thinner sections than that with LSP from one side only.

2.8 Applications of laser shock peening

Since the development of LSP, a strong interest in its commercialisation can be seen by the number of patents issued on this process. The first two key patents (Mallozi and Fairand, 1974; Clauer *et al.*, 1983) were issued in 1974 and 1983, respectively, not long after the benefits of LSP were first identified. In the period from 1996 to 2001, the General Electric Company alone received a minimum of 23 US patents based on LSP.

The increased depth of compressive residual stress produced by LSP can significantly improve properties and control the development and growth of surface cracks (Mannava and Cowie, 1996). Many of the proposed applications of LSP aim to increase fatigue life and fatigue strength of structures as well as to strengthen thin sections (Dane *et al.*, 1997; Vaccari, 1992). LSP of braze repaired (Mannava and Ferrigno, 1997) turbine components and weld repaired (Mannava *et al.*, 1997) turbine components have been patented owing to the clear improvement in properties.

A unique advantage of LSP is that the laser pulse beam can be adjusted and controlled in real time (Mannava, 1998). Through computer-controlled laser application systems, the energy per pulse can be measured and recorded for each location on the component being laser peened. If the applied laser pulse was below the specified energy, it can be redone at that time rather than after the part has failed. Regions inaccessible to SP, such as small fillets and notches, can still be treated by LSP (Vaccari, 1992; Mannava and Cowie, 1996). As long as the location can be seen, it can be laser peened (Clauer *et al.*, 1998a, b). However, no data have been found in the literature showing experimental results and possible benefits. The majority of the current applications have been the proof in principles to encourage investment for deeper research.

The spot geometry of laser beam can be changed to suit the application. A laser beam with a square profile instead of a round one allows dense, uniform packing of the laser spots. Furthermore, the process is clean and workpiece surface quality is essentially unaffected, especially for steel components. The potential of LSP includes the possibility of direct integration into manufacturing production lines with a high degree of automation (Mannava, 1998).

The aerospace industry is leading the integration of methods to apply LSP to many aerospace products, such as turbine blades and rotor components (Mannava and Cowie, 1996; Mannava and Ferrigno, 1997; Mannava *et al.*, 1997), discs, gear shafts (Ferrigno *et al.*, 2001) and bearing components (Casarcia *et al.*, 1996). LSP could also be used to treat fastener holes in aircraft skins and to refurbish fastener holes in old aircraft in which cracks, not discernible by inspection, have initiated. General Electric Aircraft Engines in the USA treated the leading edges of turbine fan blades (Mannava and Ferrigno, 1997; Mannava *et al.*, 1997) in F101-GE-102 turbine for the Rockwell B-1B bomber by LSP in 1997, which enhanced fan blade durability and resistance to foreign object damage (FOD) without harming the surface finish (Mannava and Ferrigno, 1997; Mannava *et al.*, 1997). Protection of turbine engine components against FOD (Ruschau *et al.*, 1999) is a key priority of the US Air Force. In addition, it was reported that LSP would be applied to treat engines used in the Lockheed Martin F-16C/D (Brown, 1998).

The applications of LSP can be anticipated to expand from the current field of high value, low volume parts such as hip implants and biomedical components to higher volume components such as automobile parts, industrial equipment, and tooling in the near future as high power, Q-switched laser systems become more available (Vaccari, 1992; Clauer, 1996).

2.9 Summary

The development of LSP processes has been comprehensively reviewed and addressed in this chapter. LSP has an impressive capability to improve the mechanical performance of metallic materials. The advantages and disadvantages as well as some challenges in applications of LSP can be summarised as follows.

- For potential industrial applications, laser sources for LSP require much better output performance, such as output power and repetition rate in the near future. Selection of a laser source for LSP is also required to consider two aspects: (1) the laser wavelength must be convenient in a water-confined environment; (2) the laser equipment has to be small enough to handle and easy enough to operate *in situ* in the working field.
- A high shock pressure can be generated by means of a transparent overlay and absorbent coating on a metallic material during an LSP process. Absorbent coatings can be metallic, organic paints or adhesives. The coating not only protects the metal surface from melting but also increases the magnitude of the shock pressure. The uncoated material surface can lead to very high tensile stresses, even with transparent overlay, attributed to severe surface melting. Transparent overlays can be water or glass, confining the laser energy. In order to achieve the best treatment results for individual metallic materials, the optimisation of LSP parameters, such as laser parameters, transparent overlay and absorbent coating, is very important for effective treatment.
- Peak pressure as high as 5.5 GPa with pulse duration of about 55 ns in the WCM is very useful for deep treatment of most high strength metallic materials such as titanium alloys. In WCM, the peak pressure is approximately proportional to the square root of the laser power density, when neglecting effects of the parasitic breakdown of plasma.
- LSP in a confined ablation mode can produce compressive residual stresses of about 1 mm in depth in a metallic material, which is four times deeper than that from SP. In order to generate significant compressive residual stresses below the surface of metallic materials, the optimisation of LSP parameters, such as laser parameters, must be considered in processing.
- The small laser spot can produce a higher level of surface compressive residual stresses on the material surface than the large one when using the same laser power density. But the plastically affected depth from a small impact spot could be significantly reduced.
- Although some mechanistic modelling of LSP has been conducted in the past in order to understand the dynamic process of LSP, there have been some limitations, such as unsuitable assumptions, incorrect calculations and errors in the approaches. Apparently, the finite element method as an effective analytical tool is quite an attractive technique to evaluate the dynamic stresses and the distribution of compressive residual stresses in the materials induced by LSP.

- The metallurgical physics of LSP has not been deeply investigated. Recent nano-indentation analysis of laser peened metals identified a number of phenomena previously missed that could affect the behaviour of the metal. Past research on explosively driven shock waves in metals was found to be a useful source of information to explain these phenomena.
- LSP with smaller spots with overlapping can be cost effective in practice owing to the various difficulties that exist in getting powerful lasers with pulse energies in the range of 50–200 J and a repetition rate of 1 Hz. But more systematic work is needed to address the effect of the degree of overlap, the use of planar or spherical shock waves and, in particular, the effect of gaps between laser spots, caused by laser misfire, on the resulting mechanical and metallurgical properties.
- Material type and heat treatment conditions should be also considered in process optimisation. More process modelling is also needed to understand the residual stress fields generated by LSP, in particular overlapped spots and multiple LSP in a selected area.
- A major problem for this field of research is the tremendous commercial interest in applying this technology as seen by the large number of patents produced by General Electric for turbine blade applications. Because of this commercial interest, existing basic science and process experience is either buried within the various companies or ignored because of the focus on commercially applicable empirical results. The research community's limited understanding is dangerous because if a laser peened component does fail catastrophically, what really occurred will be unknown. Few independent people will have the background able to analyse what happened let alone provide a preventative solution.

3
Simulation methodology

3.1 Introduction

Laser shock peening (LSP) is a very useful surface treatment technique in practical applications. It can create a compressive residual stress of a significant magnitude, beneath the treated surface and deep into the treated metallic components. Compressive residual stress introduced by LSP can significantly improve the mechanical performance of components, such as resistance to crack initiation and growth with extended fatigue life and enhanced fatigue strength.

Over the past 20 years, in order to improve this technique, many experimental studies on the effects of the relevant parameters of LSP have been carried out. However, dynamic responses of peened materials are very complex and it is difficult to monitor them instrumentally. To fill in this gap, a simulation technique is widely recognised as an effective tool to gain a understanding of the LSP process.

Because of the complexity of shock wave propagation in an alloy component, it is essential that the simulation can be correctly performed using a suitable computing capacity. The computer technology has developed rapidly in recent years. Two-dimensional computation with 10^6 computational cells was considered a substantial task two decades ago, but nowadays dynamic three-dimensional computation can be easily performed with more than 10^9 computational cells (Oran and Boris, 2001). This means that a normal workstation or even a desktop PC with CPU over 1 GHz and RAM over 2 Gb would have plenty of capacity to perform the simulation of complicated dynamic events like the LSP process.

A confined ablation mode applied with LSP has been demonstrated in most cases to be an effective configuration for achieving the best treatment results for metallic materials. The physical process of such typical LSP actually includes two stages. Firstly, the plasma-induced pulse pressure is generated on the material surface when a high-energy laser pulse irradiates the coating through the transparent overlay. Secondly, a residual stress field,

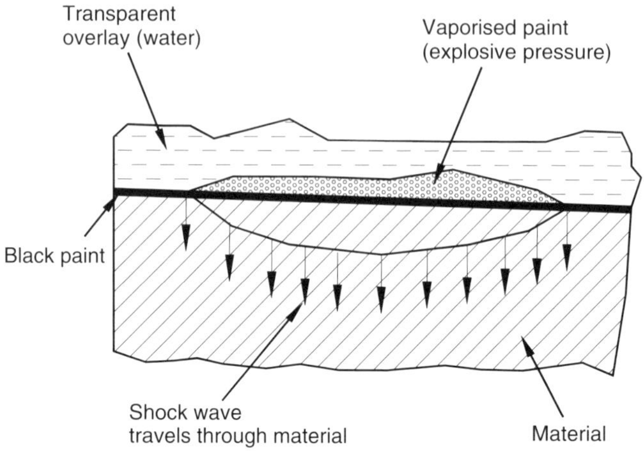

3.1 Geometry of a model in the confined ablation mode.

caused by shock waves driven by the pulse pressure, is created in the material. For the first stage, the pulse pressure can be estimated using laser physics performed by Fabbro *et al.* (1990).

One of the key interests in characterising LSP is to simulate the second stage using a mechanistic model. The aim of this chapter is to present the simulation methodology for LSP, addressing the analysis procedure.

3.2 Physics and mechanics of laser shock peening

3.2.1 Plasma pressure

When a high-energy laser pulse is focused onto a metal surface, passing through a transparent overlay and striking the opaque overlay of a surface, the heated zone on the surface is vaporised, reaching temperatures up to 10 000 °C and then is transformed into plasma by ionisation. Blowing off the high-temperature plasma on the surface can induce a high pulse pressure on the material surface. As a result, shock waves are produced inside the material. This process is depicted in Fig. 3.1.

A physical model to predict the pulse pressure as a function of laser power density was established by Fabbro *et al.* (1990), characterising the difference between confined plasma and freely expanding plasma. The configuration of the model is shown in Fig. 3.2. According to the evaluation (Fabbro *et al.*, 1990), pulse pressure, P, thickness, L, between the material surface and the overlay and expansion velocity, V, of the plasma during laser irradiation are calculated as a function of time, t, using:

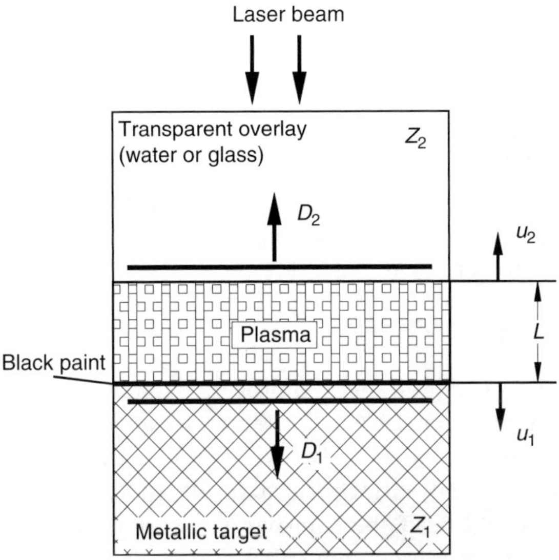

3.2 Geometry of a model in the confined ablation mode (Fabbro *et al.*, 1990).

$$I(t) = P(t)\frac{dL(t)}{dt} + \frac{3}{2\alpha}\frac{d}{dt}[P(t)L(t)] \qquad [3.1]$$

$$V(t) = \frac{dL(t)}{dt} = \left(\frac{1}{Z_1} + \frac{1}{Z_2}\right)P(t) \qquad [3.2]$$

where I is the laser power density. The shock impedance Z_i is defined as $Z_i = \rho_i D_i$, where ρ_i and D_i are the material density and the shock velocity, respectively. The index, i, represents the different materials.

In the case of a constant laser power density, I_0, the scaling law for the pulse pressure can be estimated (Fabbro *et al.*, 1990; Ballard *et al.*, 1991; Peyre *et al.*, 1996) by:

$$P(\text{GPa}) = 0.01\sqrt{\frac{\alpha}{2\alpha + 3}}\sqrt{Z(\text{g/cm}^2\,\text{s})}\sqrt{I_0(\text{GW/cm}^2)} \qquad [3.3]$$

where P is the peak pressure and α is the efficiency of the interaction. During the interaction, the total energy, E_T, from the laser source is converted into two parts. One part of the energy, αE, contributes to establishing the pulse pressure, while the other part of the energy, $(1 - \alpha)E$, is devoted to the generation and ionisation of plasma. $\alpha = 0.2$–0.5, typically (Clauer and Lahrman, 2001). Z is the combined shock impedance defined by:

$$\frac{2}{Z} = \frac{1}{Z_1} + \frac{1}{Z_2} \qquad [3.4]$$

where Z_1 and Z_2 are the shock impedance of the material and the confining overlay, respectively. If a confined ablation mode with water overlay is applied in the process, the equation [3.3] can be simplified (Peyre *et al.*, 1996) as:

$$P(\text{GPa}) = 1.02\sqrt{I_0(\text{GW/cm}^2)} \quad [3.5]$$

Equation [3.5] is readily applied in evaluation. Thus, for a water-confined ablation mode, the peak of pulse pressure is approximately the square root of the incident laser power density.

3.2.2 Shock wave

Actually, an LSP process is not a thermal process but a mechanical process for treating materials (Clauer, 1996). The thermal vapour and the plasma generated by a high-energy laser pulse are confined by the transparent overlay against the material surface. As a result, the pressure is raised much more than it would be if the transparent overlay were absent. For example, in the case of a water-confined ablation mode, it has been shown that the magnitude of the shock wave was increased by more than one order of magnitude. The plasma pressure, P, can reach 5–6 GPa, and the duration of the pulse pressure was two or three times longer than that for a direct ablation mode (Peyre *et al.*, 1998a). The shock waves resulting from expansion of high-pressure plasma are generated simultaneously, propagating into the material.

Equation [3.3] implies that the pulse pressure depends strongly on the laser power density and the impedance properties of the medium surrounding the interaction zone. Low impedance coatings on the surface can result in an increase in the pulse pressure when transmitting the shock wave into the substrate (Peyre *et al.*, 1998a). Shock wave mechanics are well presented in many textbooks (e.g. Kolsky, 1953; Nowacki, 1978; Achenbach, 1973). During propagation of a shock wave in a material, if the peak stress is above the dynamic yield strength of the material, the material yields and deforms plastically. As the shock wave is dissipated in the material, the peak stress of the shock wave decreases, but the deformation of material continues until the peak stress attenuates below the dynamic yield strength. The plastic deformation caused by shock waves leads to strain hardening and residual stresses in the material.

3.3 Mechanical behaviour of materials

3.3.1 Elastic stress–strain relation

When considering deformation of a material under applied loads, the stress–strain relationship governing behaviour of an isotropic elastic solid can be expressed by the 'generalised Hooke's law' (Kolsky, 1953):

$$\begin{aligned}
\sigma_{xx} &= \lambda\Delta + 2\mu\varepsilon_{xx} \\
\sigma_{yy} &= \lambda\Delta + 2\mu\varepsilon_{yy} \\
\sigma_{zz} &= \lambda\Delta + 2\mu\varepsilon_{zz} \\
\tau_{xy} &= \mu\varepsilon_{xy} \\
\tau_{yz} &= \mu\varepsilon_{yz} \\
\tau_{zx} &= \mu\varepsilon_{zx}
\end{aligned} \qquad [3.6]$$

where the dilatation ($\Delta = \varepsilon_{xx} + \varepsilon_{yy} + \varepsilon_{yy}$) is the change in volume of a unit cube and the two elastic constants, λ and μ, are Lamé's constants. In terms of Young's modulus, E, Poisson's ratio, ν, and shear modulus, G, the two constants can be defined as (Johnson, 1972):

$$\mu = G = \frac{E}{2(1+\nu)} \qquad [3.7]$$

$$\lambda = \frac{E\nu}{(1+\nu)(1-2\nu)} \qquad [3.8]$$

3.3.2 Von Mises yield criterion

Laser-peened targets usually are metallic materials. Metallic materials usually obey the Von Mises yielding criterion (Hoffman and Scahs, 1953):

$$\sigma_{eq} = Y^0 \qquad [3.9]$$

where Y^0 is the yield stress and σ_{eq} is the equivalent stress, defined by:

$$\sigma_{eq} = \frac{1}{\sqrt{2}}\sqrt{(\sigma_1-\sigma_2)^2+(\sigma_2-\sigma_3)^2+(\sigma_3-\sigma_1)^2}$$

or $\qquad [3.10]$

$$\sigma_{eq} = \sqrt{\frac{3}{2}\mathbf{S}:\mathbf{S}}$$

where $\mathbf{S}$ is the deviatoric stress tensor and σ_1, σ_2 and σ_3 are the three principal stresses.

3.3.3 Hugoniot elastic limit

Stress waves transmitted through an elastic–plastic material can be separated into two distinct waves, an elastic wave with a magnitude in the Hugoniot elastic limit (HEL) and a plastic wave. The magnitude of HEL depends on the material properties, defined by Johnson and Rohde (1971):

$$\text{HEL} = \frac{\lambda+2\mu}{2\mu}\sigma_Y^{dyn} = \frac{(1-\nu)\sigma_Y^{dyn}}{(1-2\nu)} \qquad [3.11]$$

where σ_Y^{dyn} is the dynamic yield strength at a high strain rate (about $10^6 s^{-1}$) and ν is Poisson's ratio.

Peyre *et al.* (1998b) conducted experiments using a Doppler-laser velocimeter system (VISAR) to determine HEL of metallic materials such as 55C1 steel alloy and 316L stainless steel. The estimated σ_Y^{dyn} for both steels was compared with their normal yield strength in the quasi-static state ($10^{-3}s^{-1}$). It was found that, for both alloys, the estimated σ_Y^{dyn} at $10^6 s^{-1}$ was more than a factor of two higher than the yield strength (σ_Y) measured at $10^{-3}s^{-1}$. Thus, in an LSP process, the dynamic yield strength at the high strain rate of $10^6 s^{-1}$ must be taken into account in analyses.

3.3.4 Elastic–plastic stress–strain relation

In an elastic solid material, the stress–strain relation is a linear function according to the Hooke's law, but for an elastic–plastic solid material, the relation is generally non-linear. The increment of total strain tensor can be decomposed into elastic and plastic components:

$$d\varepsilon = d\varepsilon^e + d\varepsilon^p \qquad [3.12]$$

where $d\varepsilon^e$ is the incremental elastic strain tensor and $d\varepsilon^p$ is the incremental plastic strain tensor.

The plastic strain increment (Prandtl–Reuss equations) (Desai and Siriwardane, 1984) can be determined by:

$$\begin{aligned}
d\varepsilon_x^p &= \frac{2}{3}d\lambda\left[\sigma_x - \frac{1}{2}(\sigma_y + \sigma_z)\right]\\
d\varepsilon_y^p &= \frac{2}{3}d\lambda\left[\sigma_y - \frac{1}{2}(\sigma_z + \sigma_x)\right]\\
d\varepsilon_z^p &= \frac{2}{3}d\lambda\left[\sigma_z - \frac{1}{2}(\sigma_x + \sigma_y)\right] \qquad [3.13]\\
d\varepsilon_{xy}^p &= d\lambda\tau_{xy}\\
d\varepsilon_{yz}^p &= d\lambda\tau_{yz}\\
d\varepsilon_{zx}^p &= d\lambda\tau_{zx}
\end{aligned}$$

where $d\lambda$ is a non-negative scalar factor, which may vary throughout the loading history.

The definition of plastic flow behaviour of materials is important in developing a plastic stress–strain relation. Plastic flow occurs when the stress state reaches the yield criterion, *f*. According to the theory of plasticity (Kachanov, 1971), the direction of plastic strain vectors is defined using a flow rule by assuming the existence of a plastic potential function, to which the incremental strain vectors are orthogonal. The increments of plastic strain can be expressed as:

$$d\varepsilon_{ij}^{p} = d\lambda \frac{\partial Q}{\partial \sigma_{ij}} \quad [3.14]$$

where Q is the plastic potential function and $d\lambda$ ($d\lambda = Gdf$) is a positive scalar factor of proportionality. For many materials, the plastic potential function, Q, and the yield function, f, can be assumed to be the same. Such materials are considered to follow the associative flow rule of plasticity. However, for geologic materials, the yield function, f, and the plastic potential function, Q, are often different from each other. These materials are considered to follow non-associative flow rules of plasticity. However, for metallic materials, the plastic potential function, Q, can be assumed to be the same as the yield function, f, while the yield function is normally assumed to be the Von Mises criterion. In terms of the equations [3.9] and [3.10], it can be expressed as:

$$f = \sigma_{eq} - Y^{0} \quad [3.15]$$

3.4 Analytical modelling

3.4.1 Basic equations of stress wave

In the theory of elasticity, a solid body is considered to be in static equilibrium under the quasi-static (or gradual) action of an external force. Its corresponding elastic deformation is assumed instantly. These assumptions are sufficiently accurate for problems in which the application of forces and the establishment of effective equilibrium are completed instantly. However, when the forces are applied on the body for a very short period of time, or when they change suddenly, effective equilibrium cannot be established instantly and the propagation of a stress wave in the body must be considered. There are two types of elastic wave in an isotropic solid, namely a dilatation wave and a distortion wave. When a solid is deformed under an applied force, both distortional and dilatational waves will normally be generated. When the stress wave propagates through the solid, internal dissipation and mechanical relaxation of the material will lead to its attenuation. Two other types of stress waves, namely a shock wave (Nowacki, 1978) and a plastic wave (Kolsky, 1953), can occur in a solid in which the stress–strain relation has ceased to be linear.

Particle motion in a plane dilatational wave is along the direction of propagation, whilst in a plane distortional wave it is perpendicular to the direction of propagation. Neglecting body forces, basic equations of motion corresponding to the two types of stress waves (dilatational and distortional) are given as follows in a Cartesian coordinate system of x–y–z (Kolsky, 1953):

$$\begin{cases} \rho \dfrac{\partial^2 u}{\partial t^2} = (\lambda + \mu)\dfrac{\partial \Delta}{\partial x} + \mu \nabla^2 u \\ \rho \dfrac{\partial^2 v}{\partial t^2} = (\lambda + \mu)\dfrac{\partial \Delta}{\partial y} + \mu \nabla^2 v \\ \rho \dfrac{\partial^2 w}{\partial t^2} = (\lambda + \mu)\dfrac{\partial \Delta}{\partial z} + \mu \nabla^2 w \end{cases} \qquad [3.16]$$

where u, v and w are the displacements in the directions, x, y and z, respectively, λ and μ are Lamé's constants and ∇^2 is the Laplace operator defined by:

$$\nabla^2 = \frac{\partial^2}{\partial x^2} + \frac{\partial^2}{\partial y^2} + \frac{\partial^2}{\partial z^2} \qquad [3.17]$$

and Δ is the first strain invariant that can alternatively be expressed by displacements as:

$$\Delta = \frac{\partial u}{\partial x} + \frac{\partial v}{\partial y} + \frac{\partial w}{\partial z} \qquad [3.18]$$

In equation [3.16], if both sides of the first equation are differentiated with respect to x, the second equation with respect to y and the third equation with respect to z, adding them, equation [3.16] can be expressed as:

$$\rho \frac{\partial^2 \Delta}{\partial t^2} = (\lambda + 2\mu)\nabla^2 \Delta \qquad [3.19]$$

Equation [3.19] is the dilatation wave equation. It shows that the dilatation Δ propagates through the medium at a velocity of $[(\lambda + 2\mu)/\rho]^{1/2}$. While in equation [3.16], differentiating both sides of the second equation with respect to z, and the third equation with respect to y, and subtracting them, it leads to:

$$\rho \frac{\partial^2}{\partial t^2}\left(\frac{\partial w}{\partial y} - \frac{\partial v}{\partial z}\right) = \mu \nabla^2 \left(\frac{\partial w}{\partial y} - \frac{\partial v}{\partial z}\right)$$

or $\qquad [3.20]$

$$\rho \frac{\partial^2 \overline{\omega}_x}{\partial t^2} = \mu \nabla^2 \overline{\omega}_x$$

where $\overline{\omega}_x$ is the rotation about the x-axis. Similar equations may be obtained for $\overline{\omega}_y$ and $\overline{\omega}_z$. Thus the rotation wave propagates at a velocity of $(\mu/\rho)^{1/2}$.

In some cases, the polar coordinate system, r–θ–z, is more convenient for analyses because of its geometry for particular components. Thus, the equations of motion in the polar coordinate system can be expressed as:

$$\begin{cases} \dfrac{\partial\sigma_{rr}}{\partial r}+\dfrac{1}{r}\dfrac{\partial\sigma_{r\theta}}{\partial\theta}+\dfrac{\partial\sigma_{rz}}{\partial z}+\dfrac{\sigma_{rr}-\sigma_{\theta\theta}}{r}=\rho\dfrac{\partial^2 u}{\partial t^2} \\ \dfrac{\partial\sigma_{r\theta}}{\partial r}+\dfrac{1}{r}\dfrac{\partial\sigma_{\theta\theta}}{\partial\theta}+\dfrac{\partial\sigma_{\theta z}}{\partial z}+\dfrac{2\sigma_{r\theta}}{r}=\rho\dfrac{\partial^2 v}{\partial t^2} \\ \dfrac{\partial\sigma_{rz}}{\partial r}+\dfrac{1}{r}\dfrac{\partial\sigma_{\theta z}}{\partial\theta}+\dfrac{\partial\sigma_{zz}}{\partial z}+\dfrac{\sigma_{rz}}{r}=\rho\dfrac{\partial^2 w}{\partial t^2} \end{cases} \qquad [3.21]$$

3.4.2 Solutions for semi-infinite body

An analytical model to predict the residual stress induced by LSP in an elastic–plastic half space was proposed by Ballard *et al.* (1991). In the model, the physical and mechanical responses of material during LSP are described in such a way that during laser–material interaction, the pulse pressure generated by blowing off plasma crushes the treated area and creates pure uniaxial compression in the direction of shock wave propagation but tensile stretching in the plane parallel to the surface. After the interaction with surrounding zones, a compressive stress field is generated within the affected volume, while the underlying layers are in a tensile state. Based on the mechanical response, Ballard *et al.* (1991) also made some assumptions in his analytical model: (1) the shock-induced deformation is uniaxial and planar; (2) the pulse pressure is uniform in space; (3) materials obey the Von Mises yielding criterion; and (4) work hardening and viscous effects are ignored.

The analytical model is axisymmetric if a round laser spot is impacted on the centre of the material surface. A cylindrical coordinate system (r, θ, z) is used for the model; the stresses and strains in the r and θ directions along the z-axis are equivalent, based on the basic assumptions of the model. The r-axis is along the peened surface and the z-axis is along the depth of the material. As the strain is uniaxial:

$$\boldsymbol{\varepsilon}=\begin{pmatrix} 0 & 0 & 0 \\ 0 & 0 & 0 \\ 0 & 0 & \varepsilon \end{pmatrix} \qquad [3.22]$$

$$\boldsymbol{\sigma}=\begin{pmatrix} \sigma_r & 0 & 0 \\ 0 & \sigma_r & 0 \\ 0 & 0 & \sigma_z \end{pmatrix} \qquad [3.23]$$

$$\boldsymbol{\varepsilon}_{\mathrm{p}}=\begin{pmatrix} -\varepsilon_{\mathrm{p}}/2 & 0 & 0 \\ 0 & -\varepsilon_{\mathrm{p}}/2 & 0 \\ 0 & 0 & \varepsilon_{\mathrm{p}} \end{pmatrix} \qquad [3.24]$$

where $\boldsymbol{\varepsilon}$, $\boldsymbol{\sigma}$, $\boldsymbol{\varepsilon}_{\mathrm{p}}$ are the strain, stress and plastic strain tensors, respectively.

The model is for a one-dimensional problem of wave propagation and the constitutive equations from the generalised Hooke's law, equation [3.6], can be simplified as:

$$\sigma_z = (\lambda + 2\mu)\varepsilon, \; \sigma_r = \lambda\varepsilon \qquad \text{(elastic)} \qquad [3.25]$$

$$\sigma_z = (\lambda + 2\mu)\varepsilon - 2\mu\varepsilon_p, \; \sigma_r = \lambda\varepsilon + \mu\varepsilon_p \qquad \text{(elastic–plastic)} \qquad [3.26]$$

As $\sigma_{rr} = \sigma_{\theta\theta} = \sigma_r$, $\sigma_{zz} = \sigma_z$, $\sigma_{r\theta} = \sigma_{rz} = \sigma_{\theta z} = 0$ and $u = v = 0$, the equation of motion from equation [3.21] and the equation of continuity can be simplified as:

$$\frac{\partial\sigma_z}{\partial z} - \rho\frac{\partial^2 w}{\partial t^2} = \frac{\partial\sigma_z}{\partial z} - \rho\frac{\partial v_z}{\partial t} = 0 \qquad [3.27]$$

$$\frac{\partial\varepsilon}{\partial t} = \frac{\partial v_z}{\partial z} \qquad [3.28]$$

where v_z is the material velocity in the z direction and ρ is the material density.

If the stress wave transmitting through the space does not induce any yielding of material as defined by the Von Mises yielding criterion ($|\sigma_r - \sigma_z| < Y^0$), the constitutive equations [3.25], substituted by equations [3.27] and [3.28], can be written as:

$$\begin{cases} (\lambda + 2\mu)\dfrac{\partial v_z}{\partial z} - \dfrac{\partial\sigma_z}{\partial t} = 0 \\ \lambda\dfrac{\partial v_z}{\partial z} - \dfrac{\partial\sigma_r}{\partial t} = 0 \end{cases} \qquad [3.29]$$

If the magnitude of stress wave is high enough to cause yielding of material (i.e. $|\sigma_r - \sigma_z| \geq Y^0$), the equations of motion and continuity are still the same as equations [3.27] and [3.28], but the constitutive equations [3.26], by eliminating the plastic strain (ε^p), substituted by equations [3.27] and [3.28], can be expressed as:

$$\begin{cases} \left(\lambda + \dfrac{2\mu}{3}\right)\dfrac{\partial v_z}{\partial z} - \dfrac{\partial\sigma_z}{\partial t} = 0 \\ \dfrac{\partial\sigma_r}{\partial t} - \dfrac{\partial\sigma_z}{\partial t} = 0 \end{cases} \qquad [3.30]$$

Plastically affected depth

The plastic strain depends only on the depth, z, in such a one-dimensional model. The absolute value of the plastic strain is a decreasing function of

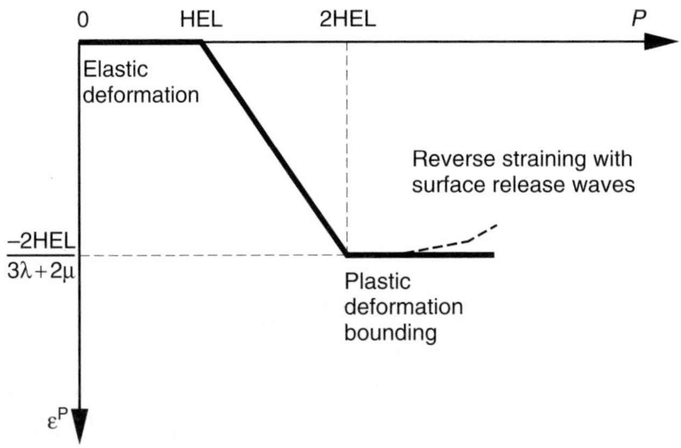

3.3 Plastic strain induced by LSP as a function of peak pressure (Peyre *et al.*, 1996).

z. On the surface of the material, ignoring work hardening and viscous effects, the plastic strain, ε_p, depends only on the magnitude of the pulse pressure, P (Fig. 3.3).

The surface plastic strain, ε_p, can be written as (Ballard *et al.*, 1991; Peyre *et al.*, 1996):

$$\varepsilon_p = \frac{-2\text{HEL}}{3\lambda + 2\mu}\left(\frac{P}{\text{HEL}} - 1\right) \qquad [3.31]$$

where HEL is the Hugoniot elastic limit, P is the peak of pulse pressure and λ and μ are the Lamé's constant. In equation [3.31], it is assumed that the pressure increases linearly between $1 \times \text{HEL}$ and $2 \times \text{HEL}$. If P is equal to $2 \times \text{HEL}$, the surface plastic strain is saturated (Fig. 3.3). But, if P is greater than $2 \times \text{HEL}$ and less than $2.5 \times \text{HEL}$, no further plastic deformation occurs (Ballard *et al.*, 1991; Peyre *et al.*, 1996).

In order to determine the residual stress field in the material, the plastically affected depth, L_p, for any given shock condition must be defined *a priori*. Ballard *et al.* (1991) found that L_p is a function of temporal profile of pulse pressure using the characteristic diagram of stresses. The estimation of L_p is expressed as:

$$L_p = \left(\frac{C_{el}C_{pl}\tau}{C_{el} - C_{pl}}\right)\left(\frac{P - \text{HEL}}{2\text{HEL}}\right) \qquad [3.32]$$

where C_{el} and C_{pl} are the elastic and plastic wave velocities, respectively, in the material, τ is the duration of pulse pressure. C_{el} can be defined from equations [3.27] and [3.29]:

$$\frac{dz}{dt} = C_{el} = \sqrt{\frac{\lambda + 2\mu}{\rho}} \qquad [3.33]$$

while C_{pl} can be defined from equations [3.27] and [3.30]:

$$\frac{dz}{dt} = C_{pl} = \sqrt{\frac{\lambda + 2\mu/3}{\rho}} \qquad [3.34]$$

where ρ is the density of material.

Residual stress

Ballard *et al.* (1991) found that the residual stress field induced by LSP can be determined through the characteristic diagrams of stresses and material velocities. Given the time, $t = +\infty$, after the impact, the residual stress field can be deduced by those characteristic diagrams. When the plastic strain and the plastically affected depth are both known, the surface residual stress with a square laser spot can be expressed (Fabbro *et al.*, 1998; Peyre *et al.*, 1998c) as:

$$\sigma_{surf} = \sigma_0 - [\mu\varepsilon_p(1+\nu)/(1-\nu) + \sigma_0]\left[1 - \frac{4\sqrt{2}}{\pi}(1+\nu)\frac{L_p}{a}\right] \qquad [3.35]$$

where a is the edge of a square laser spot and σ_0 is the initial residual stress.

If a circular laser spot of radius 'r' impacts on the material surface using $r\sqrt{2}$ instead of a in equation [3.35], the surface residual stress can be expressed (Fabbro *et al.*, 1998; Peyre *et al.*, 1998c) as:

$$\sigma_{surf} = \sigma_0 - [\mu\varepsilon_p(1+\nu)/(1-\nu) + \sigma_0]\left[1 - \frac{4\sqrt{2}}{\pi}(1+\nu)\frac{L_p}{r\sqrt{2}}\right] \qquad [3.36]$$

where σ_0 is the initial residual stress.

3.5 Finite element modelling for laser shock peening

Finite element modelling (FEM) is widely used as a powerful numerical tool for analyses of many engineering mechanics problems. FEM is able to deal with complex configurations and diverse material behaviour in practical situations.

In an LSP process, the plasma pressure generated by the laser pulse on the surface of material lasts in the order of 50–100 ns when the pressure can exceed twice the dynamic yield strength of the material. The high magnitude shock wave results in plastic deformation and favourable compressive residual stresses under the surface. Governing equations involved in the FE algorithms to solve this type of transient dynamic case have been well established in the past and can be solved through the use of developed computational techniques (Al-Obaid, 1990, 1991). It is clear that each FE solution solves a specific model using a particular algorithm and all inputs for the

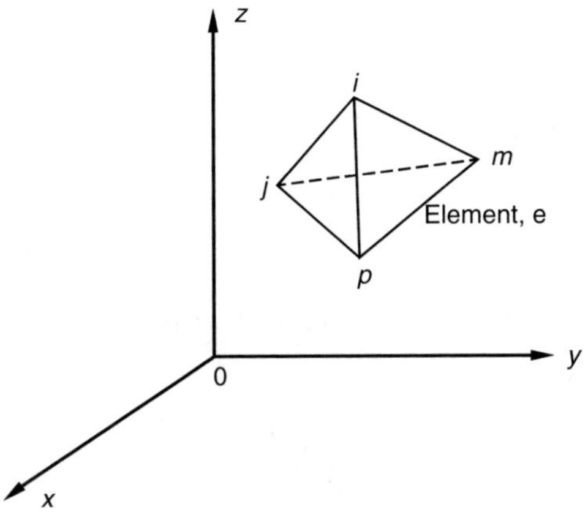

3.4 A tetrahedral element, e, with nodes *i*, *j*, *m* and *p*.

analysis can be reflected in the predicted responses. The fundamental concepts of FEM, as well as the simulation procedure of LSP using FEM, are briefly presented and discussed in the following sections, based on a commercial FEM package, ABAQUS/Explicit and ABAQUS/Standard (ABAQUS, 1998).

3.5.1 Introduction to finite element modelling

Displacement matrix

In order to state the fundamental concepts of FEM, a typical tetrahedral finite element of four nodes, e, is presented as an example in this section. This element is defined by nodes, i, j, m, p, in the space defined by the x, y and z coordinates, shown in Fig. 3.4. The state of displacement of a point is defined by three displacement components, u, v and w, in the directions of three coordinates x, y and z, respectively. The displacement vector, $\mathbf{u}$, can be expressed as:

$$\mathbf{u} = \begin{Bmatrix} u \\ v \\ w \end{Bmatrix} \qquad [3.37]$$

The element displacement is defined by a total of twelve displacement components of four nodes as:

$$\mathbf{a}^{\mathrm{e}} = \begin{Bmatrix} \mathbf{a}_i \\ \mathbf{a}_j \\ \mathbf{a}_m \\ \mathbf{a}_p \end{Bmatrix} \qquad [3.38]$$

with

$$\mathbf{a}_k = \begin{Bmatrix} u_k \\ v_k \\ w_k \end{Bmatrix}, \quad k = i, j, m, p$$

where $\mathbf{a}^e$ is a displacement vector of element e and $\mathbf{a}_k$ is a displacement vector at anode.

The displacements of an arbitrary point can be written as:

$$\mathbf{u} = \mathbf{N}\mathbf{a}^e = [\mathbf{I}N_i, \mathbf{I}N_j, \mathbf{I}N_m, \mathbf{I}N_p]\mathbf{a}^e \qquad [3.39]$$

where the matrix **N** is the shape function and I is a three-by-three identity matrix.

Strain matrix

In an elastic continuum, with displacements known at all points within the element, strains at any point can be expressed as:

$$\boldsymbol{\varepsilon} = \mathbf{B}\mathbf{a}^e \qquad [3.40]$$

where $\boldsymbol{\varepsilon}$ is the vector of strain components for the element and **B** is the strain–displacement matrix. **B** is determined by the shape functions of equation [3.39].

$$\mathbf{B} = \begin{bmatrix} N_{ix} & 0 & 0 & N_{jx} & 0 & 0 & \cdots \\ 0 & N_{iy} & 0 & 0 & N_{jy} & 0 & \cdots \\ 0 & 0 & N_{iz} & 0 & 0 & N_{jz} & \cdots \\ 0 & N_{iz} & N_{iy} & 0 & N_{jz} & N_{jy} & \cdots \\ N_{iz} & 0 & N_{ix} & N_{jz} & 0 & N_{jx} & \cdots \\ N_{iy} & N_{ix} & 0 & N_{jy} & N_{jx} & 0 & \cdots \end{bmatrix} \qquad [3.41]$$

with

$$N_{kx} = \frac{\partial N_k}{\partial x} \quad k = i, j, m, p$$

$$N_{ky} = \frac{\partial N_k}{\partial y} \quad k = i, j, m, p$$

$$N_{kz} = \frac{\partial N_k}{\partial z} \quad k = i, j, m, p$$

Stress matrix

In an elastic continuum, the relationship between stresses and strains from equation [3.6] can be written as:

$$\boldsymbol{\sigma} = \mathbf{D}(\boldsymbol{\varepsilon} - \varepsilon_0) + \boldsymbol{\sigma}_0 \qquad [3.42]$$

where $\boldsymbol{\sigma}_0$ is the initial residual stresses, $\boldsymbol{\varepsilon}_0$ is the initial strains and **D** is the matrix of elastic moduli containing appropriate material properties. **D**, in terms of the usual elastic constants E (Young's modulus) and ν (Poisson's ratio), can be written as:

$$\mathbf{D} = \begin{bmatrix} \lambda+2G & \lambda & \lambda & 0 & 0 & 0 \\ \lambda & \lambda+2G & \lambda & 0 & 0 & 0 \\ \lambda & \lambda & \lambda+2G & 0 & 0 & 0 \\ 0 & 0 & 0 & G & 0 & 0 \\ 0 & 0 & 0 & 0 & G & 0 \\ 0 & 0 & 0 & 0 & 0 & G \end{bmatrix} \qquad [3.43]$$

For an elastic–perfectly plastic material in the absence of hardening or softening, it is assumed that once a stress state reaches a failure surface f, subsequent changes in stress may shift the stress state to a different position on the failure surface, but not outside it, thus:

$$\frac{\partial f}{\partial \boldsymbol{\sigma}} \mathrm{d}\boldsymbol{\sigma} = 0 \qquad [3.44]$$

A basic assumption made in establishing the stress–strain relations for elastic–perfectly plastic materials is that for each load increment the corresponding strain increment can be decomposed into elastic and plastic components, as shown in equation [3.12]. Assuming the stress changes are generated by elastic strain components only, the stress increment is expressed by substituting equations [3.12] and [3.14] into [3.42]:

$$\mathrm{d}\boldsymbol{\sigma} = \mathbf{D}^{\mathrm{e}} \left(\mathrm{d}\boldsymbol{\varepsilon} - \lambda \frac{\partial Q}{\partial \boldsymbol{\sigma}} \right) \qquad [3.45]$$

Substituting equation [3.45] into [3.44] leads to:

$$\mathbf{D}^{\mathrm{p}} = \frac{\mathbf{D}^{\mathrm{e}} \dfrac{\partial Q}{\partial \boldsymbol{\sigma}} \left(\dfrac{\partial f}{\partial \boldsymbol{\sigma}} \right)^{\mathrm{T}} \mathbf{D}^{\mathrm{e}}}{\left(\dfrac{\partial f}{\partial \boldsymbol{\sigma}} \right)^{\mathrm{T}} \mathbf{D}^{\mathrm{e}} \dfrac{\partial Q}{\partial \boldsymbol{\sigma}}} \qquad [3.46]$$

where T refers to the transformation.

Explicit versions of $\mathbf{D}^{\mathrm{p}}$ may be obtained for some basic potential functions such as the Von Mises yielding criterion (Zienkiewicz, 1977). The final relationship between the increments of stress and increments of strain is described by:

$$d\boldsymbol{\sigma} = (\mathbf{D}^e - \mathbf{D}^p)d\boldsymbol{\varepsilon} \tag{3.47}$$

where $\mathbf{D}^e$ is the elastic matrix and $\mathbf{D}^p$ is the plastic matrix.

Stiffness matrix

The internal virtual work, U, for the element, e, associated with equations [3.40] and [3.42] can be written as:

$$U^e = \int_e \delta\varepsilon^{e^T}\sigma^e dV^e = \int_e \delta\mathbf{a}^{e^T}\mathbf{B}^{e^T}\mathbf{D}^e\mathbf{B}^e\mathbf{a}^e dV^e = \delta\mathbf{a}^{e^T}\mathbf{k}^e\mathbf{a}^e \tag{3.48}$$

where

$$\mathbf{k}^e = \int_e \mathbf{B}^{e^T}\mathbf{D}^e\mathbf{B}^e dV^e = \mathbf{B}^{e^T}\mathbf{D}^e\mathbf{B}^e V^e$$

$\mathbf{k}^e$ is the stiffness matrix and V^e is the volume of element e.

Finite element equations

For a linear elastic problem, the finite element (FE) equations are established on the principle of virtual work, which are entirely equivalent to those of internal and boundary equilibrium, strain–displacement compatibility and constraint conditions. For a body occupying a region V and having a surface S, the principle of virtual work may be written (Zienkiewicz *et al.*, 1991; Bathe, 1996) as:

$$\int_V \delta\boldsymbol{\varepsilon}^T\boldsymbol{\sigma}dV = \int_S (\delta\mathbf{u}^T\mathbf{T})dS + \int_V \delta\mathbf{u}^T\gamma dV \tag{3.49}$$

where $\boldsymbol{\sigma}$ is the stress tensor, $\mathbf{T}$ is the traction vector applied to the surface S of body, γ is the body force (unit weight) acting on the elements of body, $\delta\mathbf{u}$ is any set of virtual displacements and $\delta\boldsymbol{\varepsilon}$ is the strain tensor derived from the virtual displacements.

For a linear elastic body constituted by finite elements, the external virtual work can be expressed by the vectors of virtual nodal displacements, $\mathbf{a}$, and the nodal force $\mathbf{r}^{app}$:

$$U = \delta\mathbf{a}^T\mathbf{r}^{app} \tag{3.50}$$

While the internal virtual work of the whole body is:

$$U = \sum U^e = \sum \delta\mathbf{a}^{e^T}\mathbf{K}^e\mathbf{a}^e = \delta\mathbf{a}^T\mathbf{K}\mathbf{a} \tag{3.51}$$

where $\mathbf{K}$ is the stiffness matrix of the whole body. Then, equation [3.49] can be expressed as:

$$\mathbf{Ka} = \mathbf{r}^{\text{app}} \quad [3.52]$$

The nodal displacements, **a**, can be obtained when equation [3.52] is solved. The values of nodal displacements can be used to find the strains and stresses in any element, e, in terms of equations [3.40], [3.42] and [3.47].

3.5.2 Explicit solution procedure

An explicit time integration algorithm is adopted in the commercial ABAQUS/Explicit (ABAQUS, 1998) code. The explicit algorithm is especially well suited to solving dynamic events of high rate but short duration, such as an LSP process, which requires many small increments to obtain a high-resolution solution.

Dynamic equilibrium

A central difference rule is used to integrate the equations of motion explicitly through time, using the kinematic conditions at the next increment (ABAQUS, 1998). The dynamic equilibrium is solved at the beginning of the increment. The solution states that the nodal mass matrix, **M**, times the nodal accelerations, $\ddot{\mathbf{u}}$, equals the total nodal force s, written as:

$$\ddot{\mathbf{u}}|_{(t)} = (\mathbf{M})^{-1}(\mathbf{P} - \mathbf{I})|_{(t)} \quad [3.53]$$

where **P** is the external applied force and **I** is the internal force.

Time integration

Accelerations are integrated through time using the central difference rule, which calculates the change in velocity assuming that acceleration is constant in each increment. This change is added to the velocity at the middle of previous increment to determine that, at the middle of the current increment:

$$\dot{\mathbf{u}}\Big|_{\left(t+\frac{\Delta t}{2}\right)} = \dot{\mathbf{u}}\Big|_{\left(t-\frac{\Delta t}{2}\right)} + \frac{\left(\Delta t|_{(t+\Delta t)} + \Delta t|_{(t)}\right)}{2}\ddot{\mathbf{u}}|_{(t)} \quad [3.54]$$

Similarly, the velocities are integrated through time and added to the displacements at the beginning of the increment to determine those at the end of increment:

$$\mathbf{u}|_{(t+\Delta t)} = \mathbf{u}|_{(t)} + \Delta t|_{(t+\Delta t)}\dot{\mathbf{u}}\Big|_{\left(t+\frac{\Delta t}{2}\right)} \quad [3.55]$$

Explicit dynamic algorithm

The dynamic equilibrium at the beginning of the increment provides the acceleration (ABAQUS, 1998). When the acceleration is known, the velocity and displacement are advanced 'explicitly' through time. The term 'explicit' refers to the fact that the state at the end of increment is solely based on the exact displacements, velocities and accelerations. For a method for producing accurate results, the time increment must be quite small so that the acceleration is nearly constant during an increment of time. Since the time increments must be small, an analysis typically requires many thousands of increments. Using such an algorithm, most computational efforts are used in element calculations to determine the internal forces of elements acting on the nodes (ABAQUS, 1998), such as determining element strains and applying material constitutive relationships (the element stiffness) to determine element stresses and, consequently, the new internal forces after the increment.

The explicit algorithm can be summarised as follows:

- Nodal calculations
  a. Determine dynamic equilibrium:

  $$\ddot{\mathbf{u}}|_{(t)} = (\mathbf{M})^{-1}(\mathbf{P} - \mathbf{I})|_{(t)}$$

  b. Integrate explicitly through time:

  $$\dot{\mathbf{u}}\Big|_{\left(t+\frac{\Delta t}{2}\right)} = \dot{\mathbf{u}}\Big|_{\left(t-\frac{\Delta t}{2}\right)} + \frac{\left(\Delta t|_{(t+\Delta t)} + \Delta t|_{(t)}\right)}{2}\ddot{\mathbf{u}}|_{(t)}$$

  $$\mathbf{u}|_{(t+\Delta t)} = \mathbf{u}|_{(t)} + \Delta t|_{(t+\Delta t)}\dot{\mathbf{u}}\Big|_{\left(t+\frac{\Delta t}{2}\right)}.$$

- Element calculations
  a. Compute element strain increments, $d\varepsilon$, from the strain rate, $\dot{\varepsilon}$
  b. Compute stresses, σ, from constitutive equations:

  $$\sigma_{(t+\Delta t)} = f(\sigma_{(t)}, d\varepsilon)$$

  c. Assemble nodal internal forces, $\mathbf{I}_{(t+\Delta t)}$
- Set $t + \Delta t$ to t and return to step 1.

Stability limit

It can be expected that the time increment, Δt, has a great effect on the convergence and accuracy of results. If the time increment is larger than a critical period of time called the stability limit, Δt_{stable}, a numerical instability may lead to an unbounded solution (ABAQUS, 1998). The stability limit is defined using the highest frequency, ω_{max} of det ([**K**] – ω^2[**M**]) = 0, of the system. Without damping, the stability limit can be expressed (Cook *et al.*, 1989) as:

$$\Delta t_{stable} \leq \frac{2}{\omega_{max}} \qquad [3.56]$$

while with damping, it can be written as:

$$\Delta t_{stable} \leq \frac{2}{\omega_{max}}\left(\sqrt{1+\xi^2}-\xi\right) \qquad [3.57]$$

where ξ is the fraction of critical damping with the highest frequency.

Generally, it is not possible to determine the stability limit exactly, so conservative estimates are necessarily used in the approach. For computational efficiency, the time increment should be defined as closely as possible to the stability limit without exceeding it (ABAQUS, 1998). Unfortunately, the actual highest frequency of the system is based on a complex set of interacting factors and it is not feasible to calculate computationally its exact value. A simple estimate based on element-by-element calculation can be used, which proves efficient and conservative in practice. Using the smallest element length, L^e, and the wave speed of material, C_d, the stability limit can be estimated using (Cook *et al.*, 1989; ABAQUS, 1998):

$$\Delta t_{stable} = \frac{L^e}{C_d} \qquad [3.58]$$

with

$$C_d = \sqrt{\frac{E}{\rho}} \qquad [3.59]$$

where E is Young's modulus of material, while ρ is the mass density of material.

Comparison between explicit and implicit algorithms

For most quasi-static problems, the implicit (or standard) time integration algorithm is adopted (ABAQUS, 1998). It solves non-linear problems by means of automatic increment based on the full Newton iterative solution method (Cook *et al.*, 1989).

Both algorithms have the same dynamic equilibrium defined in terms of external applied forces, **P**, internal element forces, **I**, and the nodal accelerations, $\ddot{\mathbf{u}}$, using the same element calculations to determine the internal element forces before and after the increment. The difference between them lies in the manner in which the nodal accelerations are computed. In a non-linear problem, the standard algorithm determines the solution with iteration, but the explicit algorithm determines the solution without iteration by explicitly advancing the kinematic state from the previous one. Even though an explicit analysis can require a large number of time increments for a dynamic problem, it is more efficient than using the standard one in many cases that would require many expensive iterations (ABAQUS, 1998). In comparison, the explicit algorithm is the clear choice for wave propagation analysis, especially for a short duration transient analysis.

In addition, the explicit algorithm requires much less disc space and memory than the standard one for the same problem. For certain problems, the computational costs of the two methods may be comparable to each other, but the substantial disc space and memory savings of the explicit algorithm make it attractive in practical applications.

3.5.3 Damping

In an actual LSP process, shock waves associated with dynamic stresses dissipate and attenuate owing to damping associated with plastic deformation and material viscosity as well as wave dispersion, eventually fading away. In terms of energy dissipation in LSP, the total external work (W_t) is converted to kinetic energy (W_k), internal energy (W_i) and viscously dissipated energy (W_v), while the internal energy includes the elastically stored energy (W_e) and the plastically dissipated energy (W_p) (ABAQUS, 1998). Tracking the history of various modes of energy dissipation, in particular, plastically dissipated energy and viscously dissipated energy, would offer some insight into the evolution of the process. In terms of viscously dissipated energy, there are two typical damping models, namely bulk viscosity damping and material damping (ABAQUS, 1998).

For a dynamic process with damping like LSP, the dynamic equation [3.53] can be further improved as:

$$\mathbf{M}\ddot{\mathbf{u}} + \mathbf{C}\dot{\mathbf{u}} = \mathbf{P} - \mathbf{I} \qquad [3.60]$$

where **M** and **C** are the nodal mass and damping matrices, $\ddot{\mathbf{u}}$ and $\dot{\mathbf{u}}$ are the nodal acceleration and velocity vectors, **I** and **P** are the nodal internal force and external load vectors. Equation [3.59] can be solved using the same explicit algorithm addressed in the previous section.

In a dynamic analysis, the actual damping mechanism is usually approximated by viscous damping (Cook *et al.*, 1989). The treatment of damping in computational analyses can be categorised by two methods: phenomenological damping and spectral damping. In phenomenological damping, the models are required to describe actual physical dissipative mechanisms, such as elastic–plastic hysteresis loss, structural joint friction and material microcracking. Hence, this method has been seldom used in a practical model. There are two typical damping models in practical dynamic analysis (ABAQUS, 1998), namely viscous damping and materials damping.

Material damping

A popular spectral damping method called Rayleigh or proportional damping (also materials damping) is to form the damping matrix **C** in the dynamic equation as a linear combination of the stiffness and mass matrices:

$$\mathbf{C} = \alpha_R \mathbf{M} + \beta_R \mathbf{K} \qquad [3.61]$$

where α_R is mass proportional damping and β_R is stiffness proportional damping, respectively.

α_R defines a damping contribution proportional to the mass matrix for an element. The damping forces are caused by the absolute velocities of nodes in the model. The resulting effect can be an analogue of the model moving through a viscous fluid so that any motion of any point in the model triggers damping forces. It was found that reasonable mass proportional damping does not reduce the stability limit significantly.

However, it is normally difficult to determine whether or not α_R has adversely influenced the solution, so it is normally set to be zero (ABAQUS, 1998).

β_R defines damping proportional to the elastic material stiffness. A 'damping stress', σ_d, proportional to the total strain rate is introduced:

$$\sigma_d = \beta_R \mathbf{D}^{el} \dot{\varepsilon} \qquad [3.62]$$

where $\dot{\varepsilon}$ is the strain rate. For hyperelastic and hyperfoam materials, $\mathbf{D}^{el}$ is defined as the initial elastic stiffness. For all other materials, $\mathbf{D}^{el}$ is the current elastic stiffness of the material. This damping stress is added to the stress caused by the constitutive response at the integration point when the dynamic equilibrium equations are formed, but it is not included in the stress output. Stiffness damping can be introduced for any non-linear analysis, but it must be used with caution because it may significantly reduce the stability limit (ABAQUS, 1998).

Bulk viscosity

Material bulk viscosity normally introduces viscous damping associated with volumetric straining. Its purpose is to improve the modelling of high-speed dynamic events by limiting numerical oscillations. The ABAQUS/

Explicit algorithm contains linear and quadratic forms of bulk viscosity. Linear bulk viscosity is normally included to damp 'ringing' at the highest element frequency. It generates a bulk viscosity stress, σ_1, which is linearly proportional to the volumetric strain rate (ABAQUS, 1998):

$$\sigma_1 = b_1 \rho C_d L^e \dot{\varepsilon}_{vol} \quad [3.63]$$

where b_1 is a damping coefficient, with a default value of 0.06 (ABAQUS, 1998), ρ is the material density, C_d is the dilatational wave speed, L^e is the element characteristic length and $\dot{\varepsilon}_{vol}$ is the volumetric strain rate.

Quadratic bulk viscosity is applied only if the volumetric strain rate is compressive. The bulk viscosity stress, σ_2, is quadratic in the strain rate (ABAQUS, 1998):

$$\sigma_2 = \rho (b_2 L^e)^2 |\dot{\varepsilon}_{vol}| \min(0, \dot{\varepsilon}_{vol}) \quad [3.64]$$

where b_2 is the damping coefficient and its default value is 1.2 (ABAQUS, 1998). The quadratic bulk viscosity smears a shock front across several elements and is introduced to prevent elements from collapsing under extremely high velocity gradients.

3.6 Finite element analysis techniques

3.6.1 Integration of explicit and implicit procedures

When the explicit procedure is applied to a dynamic problem, the result may take a long time to reach convergence in static equilibrium, because of high frequency local numerical oscillation around the final converged results, even though there is damping in various forms. In this case, it can be effective to impose the standard procedure to the unsettled deformed dynamic body to achieve static equilibrium. It can be considered that such a process quickly settles the high frequency local oscillation without affecting the converged results, converting a final converging problem based on a tiny time increment to one based on Newton iteration (Meguid *et al.*, 1999a, b). Similarly, the explicit procedure can also be imposed on a deformed body in the standard procedure when it is necessary. For example, in the case of sheet metal including initial preloading, forming and subsequent springback, the initial preloading can be simulated using the standard procedure, while the subsequent forming process can be simulated using the explicit method. Finally, springback analysis can be performed with the standard procedure again. An approach integrating both explicit and standard procedures is particularly useful in simulating LSP for predicting the compressive residual stress in a metallic material.

3.6.2 Non-linear dynamic problems

The explicit procedure for solving non-linearities is usually straightforward, accurate and effective. For a non-linear problem, the explicit method requires that the internal force of each element, $\boldsymbol{r}_n^{\text{int}}$, be calculated before the new displacement $\boldsymbol{a}_{n+1}$ can be computed. Element-by-element calculation of $\boldsymbol{r}_n^{\text{int}}$ requires that element stresses, $\boldsymbol{\sigma}_n$, be known. For linear problems, $\boldsymbol{\sigma}_n = \mathbf{DBa}_n$, when $\mathbf{a}_n$ is known. For plasticity, the stress increment $d\boldsymbol{\sigma} \approx \Delta\boldsymbol{\sigma}$ can be computed from the strain increment, $\Delta\boldsymbol{\varepsilon} = \mathbf{B}(\mathbf{a}_n - \mathbf{a}_{n-1})$, and the constitutive law, equation [3.42]. Hence, the stress at time $t = n \cdot \Delta t$ is given by $\boldsymbol{\sigma}_n = \boldsymbol{\sigma}_{n-1} + \Delta\boldsymbol{\sigma}$.

For linear problems, the accuracy of an explicit solution is usually assured when the time-step stability limit, equation [3.55] or [3.56], is satisfied. This limit is also valid for non-linear problems provided that the instantaneous value of $\omega_{\max}$ is applied, which is a function of material properties, element geometry and mesh geometry (Cook *et al.*, 1989).

3.6.3 Achieving static equilibrium

When the current state of a deformed body in dynamic equilibrium in an explicit dynamic analysis is imported into a standard static analysis, the model will not initially be in static equilibrium. To achieve static equilibrium, additional artificial forces, $\mathbf{r}^{\text{init}}$, must be applied in the standard analysis to the deformed body in dynamic equilibrium. Such forces, $\mathbf{r}^{\text{init}}$, are balanced by both dynamic forces (inertia and damping) and boundary interaction forces. The boundary interaction forces are the result of interactions from the fixed boundary and contact conditions. Any changes in the boundary and contact conditions from the explicit analysis to the standard analysis will contribute to the forces, $\mathbf{r}^{\text{init}}$.

Such artificial forces, $\mathbf{r}^{\text{init}}$, should be vanished in static equilibrium to satisfy all original internal and boundary conditions in dynamic equilibrium. However, if the forces, $\mathbf{r}^{\text{init}}$, in the static analysis are removed instantaneously, the convergence problem will arise in the static analysis. Hence, the forces need to be removed gradually until complete static equilibrium is achieved. During this process of removing the forces, $\mathbf{r}^{\text{init}}$, the body may deform further slightly with somewhat redistributed local internal forces, leading to the final residual stress state.

In the standard procedure, the following algorithm is used to remove the forces, $\mathbf{r}^{\text{init}}$:

- The imported dynamic forces at each node point are defined at the start of the analysis as the initial forces in the materials.
- An additional set of artificial forces is introduced at each node point. These forces are equal in magnitude to the imported dynamic forces but

are opposite in sign. The sum of dynamic and artificial forces thus creates zero internal forces at each node point at the beginning of the step.

- The internal artificial forces are ramped off linearly by increments during the static analysis. A timescale from 0.0 to 1.0 can be introduced in the static analysis. The time increments are simply fractions of the timescale, based on Newton's method, adjusted automatically in the standard algorithm to ensure computational efficiency. At the end of the process artificial forces are removed completely and the remaining forces in the material define the residual stress in static equilibrium.

3.7 Laser shock peening simulation procedure

LSP can be simulated using the commercial FE package ABAQUS to determine both the short duration shock wave response and the resulting residual stress state in the material. The ABAQUS/Explicit code can be used to model the dynamic response induced by the plasma pulse pressure on the material surface, determining the dynamic stresses in the material. However, when using this code to determine the final residual stress field, the stress state is extremely slow to reach the converged state in static equilibrium even though the ABAQUS/Explicit code provides a small amount of damping in the form of bulk viscosity to control high-frequency oscillation through the whole dynamic analysis. The ABAQUS/Standard code could also be used to simulate the entire LSP process. However, the computational expense of determining the dynamic stresses is prohibitive. Therefore, it is more efficient to integrate both FE codes, with the best capacities of each, to obtain the final solution.

To simulate the residual stress field generated in a material from single LSP, the ABAQUS/Explicit code is first used to perform the dynamic analysis. Actually, full development of plastic deformation in the material during an LSP process takes much longer than the duration of the pulse pressure, owing to reflection and interaction of various shock waves. Typically, a dynamic solution should be set two orders of magnitude longer than the duration of the pulse pressure to insure that all plasticity has occurred. Since the ABAQUS/Explicit code provides a small amount of damping in the form of bulk viscosity to control high-frequency oscillation through the whole dynamic analysis, a certain solution period in the dynamic analysis can be determined when the dynamic stress state becomes steady and no further plastic deformation occurs in the material. Once the steady dynamic state is reached, the deformed body with all transient stress and strain states is imported into the ABAQUS/Standard code to determine the residual stress field at static equilibrium.

In the simulation of multiple LSP, the residual stress and strain states from the first impact become the initial stress and strain states in the mater-

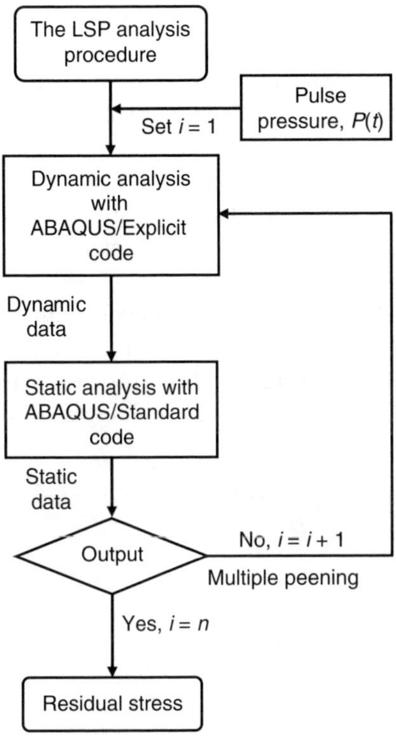

3.5 A flowchart of the LSP analysis procedure.

ial for the second impact, repeating this method for the third impact and so on. Residual stresses are obtained for each impact through the static equilibrium in the ABAQUS/Standard code when importing the steady dynamic stress and strain states from ABAQUS/Explicit. The procedures for both single and multiple LSP are summarised in Fig. 3.5.

3.8 Summary

In this chapter, the simulation methodology for LSP has been comprehensively presented and discussed. The simulation of an LSP process in a metallic material includes two stages. The first stage is for the determination of the high pulse pressure generated by laser-induced plasma, which can be defined using the approach proposed by Fabbro *et al.* (1990). The second stage is for the determination of the mechanical responses of the material peened by a plasma pulse pressure of very short duration (about 100 ns).

One of the key interests in the simulation of LSP is to evaluate and optimise the compressive residual stress in the material, which can significantly improve its mechanical performance. There are few analytical models available for analysing an LSP process because of the difficulties and complexity in a dynamic problem with a very high strain rate (at least 10^6) combined with plastic deformation and dissipation, stress wave attenuation and dispersion. This type of problem is more effectively solved by means of finite element modelling (FEM). Apparently, there are two distinct algorithms available, explicit and implicit (or standard) procedures for solving a dynamic problem, with advantages and disadvantages. An integrated approach with both algorithms is an effective and efficient procedure to simulate LSP.

The commercial FE package ABAQUS, with its specific features, allows such complicated dynamic problems to be solved with normal computational capacities. The simulation of different LSP processes using the ABAQUS/Explicit and ABAQUS/Standard codes is elaborated in the following chapters.

4

Two-dimensional simulation of single and multiple laser shock peening

4.1 Introduction

This chapter presents a two-dimensional (2D) dynamic finite element simulation of single and multiple laser shock peening (LSP) on a metallic alloy using the procedure discussed in Chapter 3. The dynamic stresses and residual stress field in the laser peened material are carefully studied and evaluated with respect to some key factors of LSP.

In the analysis, a complicated three-dimensional (3D) LSP case is simplified into a 2D one because of the geometric symmetry of the specimen and the use of a circle laser spot. In order to achieve an accurate solution, the sensitivity of a finite element mesh is carefully addressed in the selection of the model. Some factors, such as bulk viscosity and material damping, in controlling and suppressing numerical oscillations to achieve a stable dynamic stress state, are also studied and evaluated. Dynamic stresses and residual stress profiles on the surface and in the depth are presented and discussed. The predicted results are correlated with the available experimental data in the literature and estimated from the analytical model presented in Chapter 3. The residual stress in the laser peened material is studied with respect to variations in LSP conditions such as pressure, pressure duration and laser spot size.

4.2 Laser shock peening process

4.2.1 Laser shock peening conditions

The LSP process conducted by Ballard (1991) is modelled. The laser equipment used for the experiments was a Q-switched Nd-glass laser delivering an output energy of around 150 J, laser power density of 10 GW/cm^2 and a Gaussian laser pulse of a full width at half-maximum (FWHM) of about 25–30 ns. The laser pulse with a spot of 8 mm in diameter at a wavelength

Table 4.1 Material properties of 35CD4 30HRC steel alloy (Ding, 2003)

Laser peened target	35CD4 30HRC Steel
Density, ρ (kg/m^3)	7800
Elastic modulus, E (GPa)	210
Poisson's ratio, ν	0.29
Hugonoit elastic limit, HEL (GPa)	1.47
Dynamic yielding strength, σ_y^{dyn} (GPa)	0.87

of 1.064 μm produced a peak pressure of 2.8 GPa with a Gaussian temporal profile with FWHM of 50 ns.

In the experiments, the residual stress was measured by means of X-ray diffraction. As X-ray diffraction can determine the stress state only on the exposed surface of specimen, an electrolytic polishing method was used to remove successive layers of the material beneath the LSP impact spot. After a layer was removed, the newly exposed surface was measured again using X-ray diffraction.

4.2.2 Material

A 35CD4 30HRC steel alloy was selected for evaluation. It was a cylindrical solid with a diameter of 30 mm and a height of 15 mm (Ballard, 1991). In a single LSP process, the laser beam was irradiated on to the top flat surface of the specimen. The transparent overlay was a thin layer of water flowing over the specimen. The material surface was coated with black paint to prevent thermal effects. For the purpose of evaluation, the material is assumed to be homogeneous, isotropic and elastic–perfectly plastic. Material properties of the alloy are listed in Table 4.1.

4.3 Two-dimensional finite element simulation

4.3.1 Finite element model

A pulse pressure is generated on the material surface when the specimen is impacted by a single laser pulse of very short duration. The pulse pressure only lasts for a very short period (FWHM = 50 ns) resulting in a complicated shock wave into the material. As the specimen is axisymmetric in shape and the impact pressure is axisymmetric, stresses and strains are independent of the polar angle, θ, if using a cylindrical coordinate system (r, θ, y) for the model. As a result, a plane coordinate system (r, y) can be adopted with the axis, r, in the radial direction and the axis, y, in the depth direction of specimen. In this way, the model can be simplified to a 2D model for per-

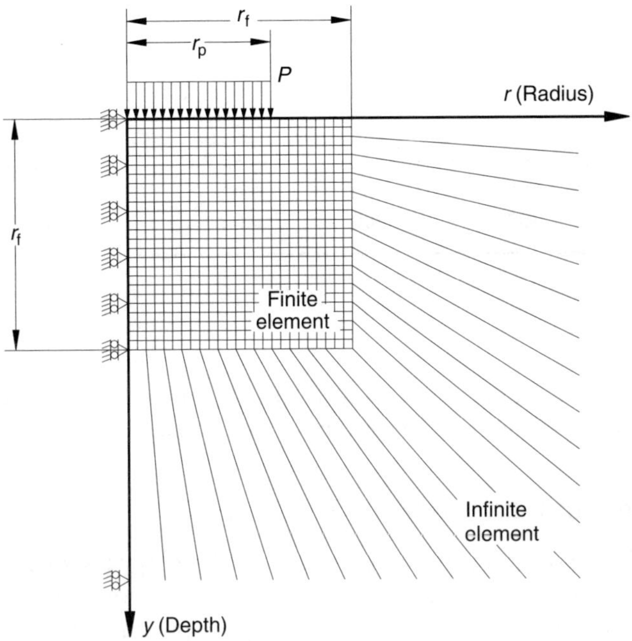

4.1 2D finite element model with axisymmetric boundary conditions.

forming finite element analyses to assure computational efficiency. A mesh configuration is demonstrated in Fig. 4.1 with four-node finite elements and infinite elements, as well as the axisymmetric boundary conditions, which are located at the centreline of the model.

The finite elements can undergo non-linearity with large deformation to cope with high-pressure impact, while the infinite elements were assumed to be elastic elements, which are used as non-reflecting boundaries for the finite element area, providing quiet boundaries that minimise the reflection of waves back into the area. If the residual stress field approaches the boundary between the finite and the infinite elements, the finite element region needs to be extended. Some preliminary analyses were conducted; eventually r_f (= 6 mm) is defined as the radius of finite element region, while r_p (= 4 mm) is the radius of the impact zone. In the analysis, the plastic yielding is defined by the Von Mises criterion.

4.3.2 Mesh refinement

In FE analyses, results are normally sensitive to the configuration of the finite element mesh. A dense mesh normally results in more accurate solutions, but with a high computational cost. In order to evaluate the effects of mesh refinement on the results, three FE models are meshed, described

Table 4.2 Configurations of three meshed FE models (Ding, 2003)

FE model	Finite element	Infinite element	Element length (L^e) (mm)	Mesh density (L^e/r_p) (%)
Ma	120 × 120	2 × 120	0.05	1.25 (coarse)
Mb	200 × 200	2 × 200	0.03	0.75 (moderate)
Mc	300 × 300	2 × 300	0.02	0.5 (fine)

in Table 4.2 as Ma, Mb and Mc, respectively. The mesh density is defined as a ratio of element length (L^e) to radius of impact zone (r_p).

4.3.3 Pressure–time history

The high plasma pressure resulting from the high-energy laser pulse impact on the material surface usually has a Gaussian temporal profile with a FWHM of 50–60 ns (Peyre *et al.*, 1998*c*). The peak pressure is dependent on the individual LSP conditions. Its peak magnitude can be approximately estimated from the pressure formula discussed in Chapter 3. In a water-confined ablation mode, the peak pressure is proportional to the square root of laser power density (Devaux *et al.*, 1993; Peyre *et al.*, 1996).

The pressure–time history is very important for simulation of an LSP process. Although the pressure–time history is usually described as a Gaussian temporal profile, it is very close to a triangular ramp because of a very narrow width of pressure pulse duration (Braisted and Brockman, 1999). Because of this, the analysis does not explicitly model the characteristics of the Gaussian pulse. The pressure–time history for the single impact or multiple impact was simplified into a triangular ramp, in which the pressure rises linearly to the peak, that is 2.8 GPa, in 50 ns and then decays linearly during the following 50 ns, shown in Fig. 4.2.

4.3.4 Analytical model

As addressed in Chapter 3, Ballard (1991) developed an analytical model to predict the surface residual stress induced by LSP in an elastic–plastic half space. The model was assumed to be axisymmetric, considering a round laser spot impacting on the centre of material surface. With considerable assumptions, the surface plastic strain, ε_p, was defined as:

$$\varepsilon_p = \frac{-2\mathrm{HEL}}{3\lambda + 2\mu}\left(\frac{P}{\mathrm{HEL}} - 1\right) \qquad [4.1]$$

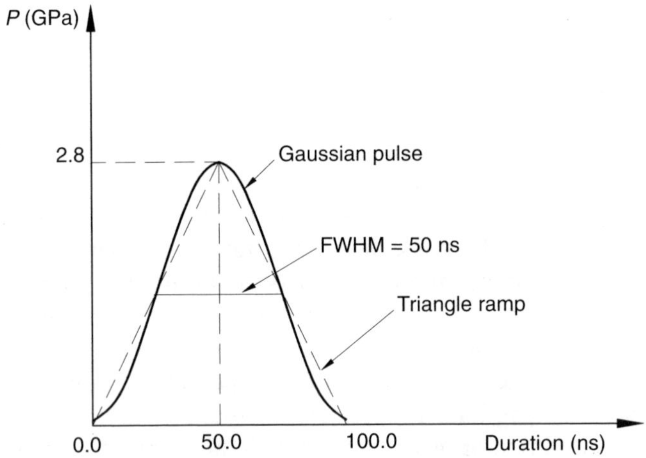

4.2 Pressure–time history of a single pressure pulse on the specimen surface.

where HEL is the Hugoniot elastic limit, P is the pressure and λ and μ are the Lamé's constants. In this case, HEL = 1.47 GPa (Table 4.1) and $P = 2.8$ Gpa (Fig. 4.2). In addition, μ and λ calculated from equations [3.7] and [3.8] in Chapter 3 are 81.4 and 112.4 GPa, respectively. Substituting HEL, P, λ and μ into equation [4.1], the surface plastic strain, ε_p, is 0.0053.

In order to determine the residual stress in the material, the plastically affected depth, L_p, for a given shock condition must be defined *a priori*. L_p can be defined by:

$$L_p = \left(\frac{C_{el} C_{pl} \tau}{C_{el} - C_{pl}} \right) \left(\frac{P - \text{HEL}}{2\text{HEL}} \right) \quad [4.2]$$

where C_{el} and C_{pl} are the elastic and plastic velocities and τ is the pressure pulse duration. In such a case, C_{el} and C_{pl}, calculated from equations [3.33] and [3.34] in Chapter 3, are 5939.8 and 4622.5 m/s, respectively. Substituting C_{el}, C_{pl}, HEL, P and τ into equation [4.2], the plastically affected depth, L_p, is 0.47 mm.

As a round laser spot with a radius of r_p (= 4 mm) was impacted on the material surface, the estimated surface residual stress was expressed as (Fabbro *et al.*, 1998; Peyre *et al.*, 1998*b*):

$$\sigma_{surf} = \sigma_0 - [\mu \varepsilon_p (1 + \nu)/(1 - \nu) + \sigma_0] \left[1 - \frac{4\sqrt{2}}{\pi} (1 + \nu) \frac{L_p}{r_p \sqrt{2}} \right] \quad [4.3]$$

where σ_0 is the initial residual stress and is set as zero in this case and ν is Poisson's ratio. Thus, the estimated surface residual stress, σ_{surf}, is about –635 MPa.

4.4 Evaluation and discussion

4.4.1 Mesh sensitivity

During a dynamic analysis using the ABAQUS/Explicit code, the time increment is always set to be less than the stability limit to avoid a numerical instability that can lead to an unbounded solution. As mentioned in Chapter 3, the stability limit is defined in terms of the highest frequency in the assembled finite element model. It was found that the highest element frequency determined on a basis of element-by-element is always higher than the highest frequency for the assembled finite element model (Cook *et al.*, 1989; ABAQUS, 1998). Therefore, the stability limit can be defined using the element length (L^e) and the wave speed (C_d) in the material. In this way, the stability limit is proportional to the smallest element dimension. For model Ma, with a uniform finite element size (L^e) of 0.05 mm and the undamped elastic wave speed of material, $C_d = 5.19 \times 10^3$ m/s, the estimated stability limit is about 9.63×10^{-9} s.

In a linear elastic material, as the wave speed (C_d) is constant, the only change in the stability limit during the analysis results from the change in the smallest element dimension. In a non-linear material, the wave speed changes as the material yields and the stiffness of material changes. After yielding, the stiffness decreases, reducing the wave speed and, consequently, increasing the stability limit (ABAQUS, 1998).

The relative computational cost as a result of mesh refinement is rather straightforward in the explicit method. Mesh refinement increases the computational cost by increasing the number of elements and reducing the smallest element dimension (i.e. stability limit). For instance, in a 2D case, if the mesh is refined by a factor of 2 in all two directions, the computational cost increases by a factor of 4 as a result of an increase in number of elements and further by a factor of 2 as a result of a decrease in the time increment (ABAQUS, 1998).

Table 4.3 shows the estimated time increment and computational costs for each model, compared with the actual ones obtained from running the analyses using ABAQUS/Explicit (on a Sun Blade 1000 workstation using two central processing units (CPUs) of 900 MHz and a RAM of 2 Gb). In the comparison, Ma is chosen as a benchmark and its time increment and CPU time are defined as *S* and *C*, respectively. After the mesh is refined to Mb and Mc, the estimated time increment is decreased to 0.6*S* for Mb and 0.4*S* for Mc, while the corresponding CPU time is increased to 4.6*C* for Mb and 15.6*C* for Mc.

In the actual dynamic analysis, the time increment in Ma was 3.9×10^{-9} s with the corresponding CPU time of 101 s. As the mesh was refined to Mb and Mc, the time increment was decreased by a factor of 0.69 for Mb and

Table 4.3 Time increment and CPU time from the simple estimate and actual running in ABAQUS/Explicit (Ding, 2003)

FE model	Simplified estimate		Actual running*	
	Δt (s)	CPU time (s)	Δt (s)	CPU time (s)
Ma	S	C	3.90×10^{-9}	101
Mb	$0.6S$	$4.6C$	2.70×10^{-9}	423
Mc	$0.4S$	$15.6C$	1.80×10^{-9}	1397

* Sun Blade 1000 Workstation with two CPUs of 900 MHz and a RAM of 2 Gb.

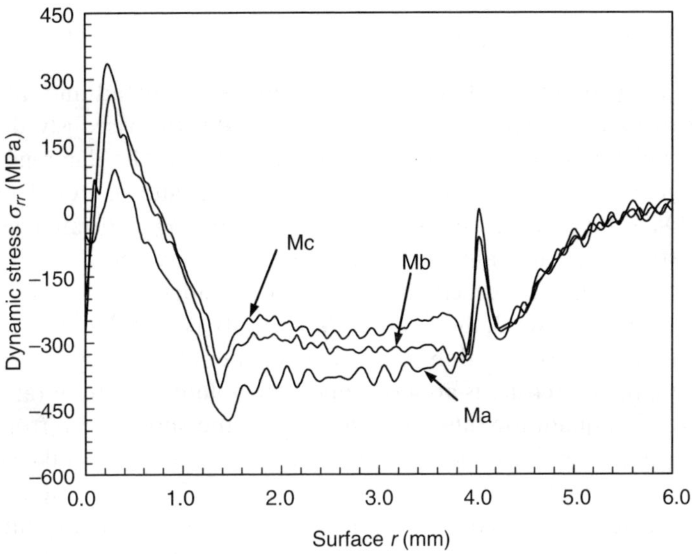

4.3 Surface dynamic stresses (σ_{rr}) from three FEA models (Ma, Mb and Mc) after 4000 ns.

0.46 for Mc, while the corresponding CPU time was increased by a factor of 4.2 for Mb and 13.8 for Mc, respectively. The simple estimation is quite well correlated with the actual time increment and CPU time with respect to the mesh refinement.

The radial stresses (σ_{rr}) on the surface of specimen affected by the mesh refinement are shown in Fig. 4.3. After a solution time of 4000 ns, the result from Mc is almost the same as that from Mb, but it differs quite significantly from Ma. Thus, in order to insure the computational efficiency, Mb was selected for further evaluation.

Table 4.4 Distribution of dissipated energy with respect to different coefficients of bulk viscosity (Ding, 2003)

Dissipated energy	BV1 (default)	BV2	BV3
	$b_1 = 0.06$ $b_2 = 1.2$	$b_1 = 0.12$ $b_2 = 2.4$	$b_1 = 0.6$ $b_2 = 12$
W_v (mJ)	180	186	217
W_p (mJ)	122	116	85
Total (mJ)	302	302	302

4.4.2 Effect of bulk viscosity

In an LSP process, shock stress waves will be dissipated and attenuated owing to damping and plasticity represented by the viscously dissipated energy (W_v) and plastically dissipated energy (W_p). The damping mechanisms mainly include bulk viscosity damping and material damping (ABAQUS, 1998). Bulk viscosity introduces damping associated with volumetric straining. There are two forms of bulk viscosity, which can be applied using the ABAQUS/Explicit code: linear and quadratic bulk viscosities. As addressed in Chapter 3, the linear bulk viscosity is included in the analysis to damp 'ringing' (oscillation) of the high element frequency. However, the quadratic bulk viscosity is applied only if the volumetric strain rate is compressive. The quadratic bulk viscosity smears the shock wave front across several elements, introduced to prevent elements from collapse under extremely high velocity gradients (ABAQUS, 1998). As addressed in Chapter 3 for bulk viscosity damping, b_1 is a damping coefficient for the linear bulk viscosity and b_2 is a damping coefficient for the quadratic bulk viscosity. By default, b_1 is 0.06 and b_2 is 1.2 in the ABAQUS/Explicit code.

To evaluate the effects of changes in the damping coefficients of bulk viscosity on the dissipated energy and the simulated results, three groups of bulk viscosity with different values of b_1 and b_2 are defined as BV1, BV2 and BV3, respectively, given in Table 4.4. Supposing that BV1 is the default, BV2 has a factor of 2 higher than BV1 whilst BV3 has a factor of 10 higher than BV1. The results from Table 4.4 show that the dissipated energy is clearly diversified by increasing the bulk viscosity damping from BV1 to BV3, after a solution time of 4000 ns. As a result of increasing the bulk viscosity damping, the viscously dissipated energy was increased to 186 mJ by 3% for BV2 and 217 mJ by 21% for BV3, while the plastically dissipated energy was decreased to 116 mJ by 5% for BV2 and 85 mJ by 30% for BV3. The total dissipated energy in the model remains unchanged at 302 mJ.

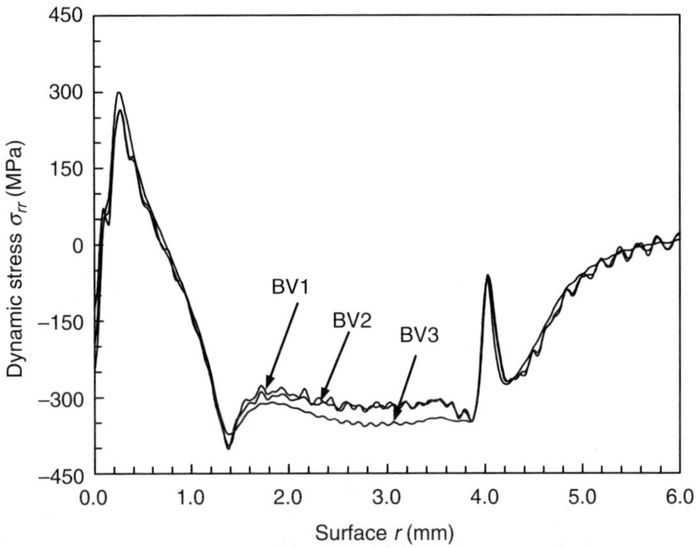

4.4 Surface dynamic stresses (σ_{rr}) resulting from three levels of bulk viscosity.

Figure 4.4 shows that the 'ringing' (or oscillation) on the surface dynamic stress (σ_{rr}) profile is gradually damped out when increasing the bulk viscosity. The surface dynamic stress profile for BV1 is very similar to that for BV2, but the 'ringing' in the surface dynamic stress profile for BV2 is slightly weaker than that for BV1. Compared with the stress profiles resulted from BV1 and BV2, the stress profile from BV3 shows the best performance in damping out 'ringing'.

In general, an increase in the bulk viscosity increases the viscously dissipated energy, consequently damping out the 'ringing' on the results. However, a significant increase in bulk viscosity may result in an unrealistic and artificial solution for a dynamic problem. Therefore, in the dynamic analyses, the coefficients of bulk viscosity were set to the default ($b_1 = 0.06$, $b_2 = 1.2$).

4.4.3 Effect of material damping

For material damping, the simulated results are evaluated through the analysis using specified values of β_R, the proportional stiffness damping. In order to avoid a dramatic drop in the time increment for the analysis, β_R should be less than, or of the same order of magnitude as, the time increment without the material damping (ABAQUS, 1998). Since the actual time increment of Mb was 2.7×10^{-9} s in Table 4.3, the value of β_R is assumed to be 1.0×10^{-9} and 2.0×10^{-9}, respectively, s for evaluation.

Table 4.5 Some considerable factors with artificially introduced values of β_R (Ding, 2003)

Considerable factors	β_R		
	$\beta_R = 0$	1.0×10^{-9}	2.0×10^{-9}
Δt (s)	2.7×10^{-9}	1.9×10^{-9}	1.4×10^{-9}
CPU time (s)	423	611	800
W_v (mJ)	180	204	220
W_p (mJ)	122	98	82
Max. σ_{rr} (MPa)	–402.0	–418.0	–506.0

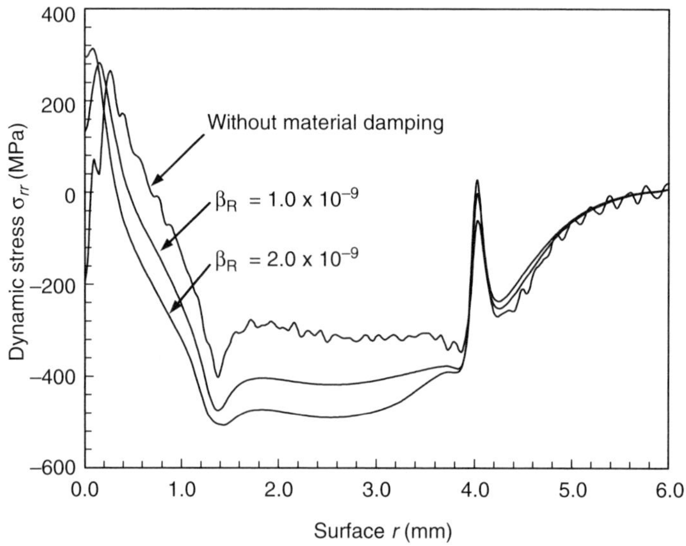

4.5 Surface dynamic stress (σ_{rr}) profiles from analyses with different material damping.

Table 4.5 shows that some key factors in the analysis are significantly affected by the values of β_R compared with those obtained from the analysis without the material damping. The increase in the value of β_R induces a reduction in the time increment, Δt, and consequently increases the CPU time. As a result of increasing the value of β_R, the viscously dissipated energy increases but the plastically dissipated energy decreases.

Figure 4.5 shows the surface dynamic stress (σ_{rr}) profiles, after a solution time of 4000 ns, resulted from the analyses with $\beta_R = 1.0 \times 10^{-9}$, $\beta_R = 2.0 \times 10^{-9}$ and no material damping, respectively. The surface stress profiles with β_R at the selected values are quite smooth, but the peak compressive stress

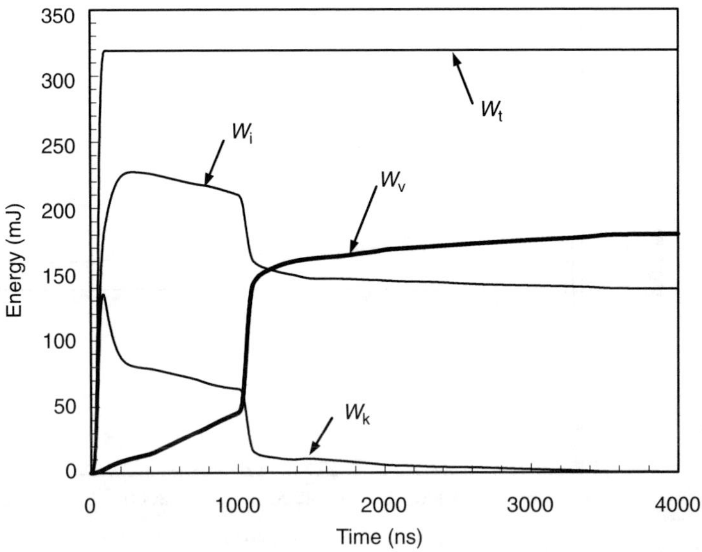

4.6 History of total external work, internal energy, kinetic energy and viscously dissipated energy in the process.

level becomes fictitious as a result of increasing the value of β_R. Compared with the peak compressive stress (402.0 MPa) obtained from the analysis without material damping, the peak compressive stress was increased to 418 MPa by 4% for the analysis with $\beta_R = 1.0 \times 10^{-9}$, and 506.0 MPa by 26% for the analysis with $\beta_R = 2.0 \times 10^{-9}$. The artificial material damping introduced in the analysis can significantly damp out numerical oscillations to achieve a smooth stress profile, but it is difficult to estimate an appropriate value of β_R in order to predict the material response in the analysis. Therefore, the following dynamic analyses do not involve artificial material damping and the viscously dissipated energy is attributed to the bulk viscosity only.

4.4.4 Energy dissipation

During a single LSP process, the total external work (W_t) of a pulse pressure on the material surface is converted to the kinetic energy (W_k), the internal energy (W_i) and the viscously dissipated energy (W_v). The history of these energies is plotted in Fig. 4.6. After releasing the pulse pressure (about 100 ns), the total external work of 320 mJ is converted to the kinetic energy, the internal energy and the viscously dissipated energy, respectively. After 1000 ns, the kinetic energy and internal energy dramatically decrease, gradually approaching 0 mJ and 130 mJ, respectively, while the viscously dis-

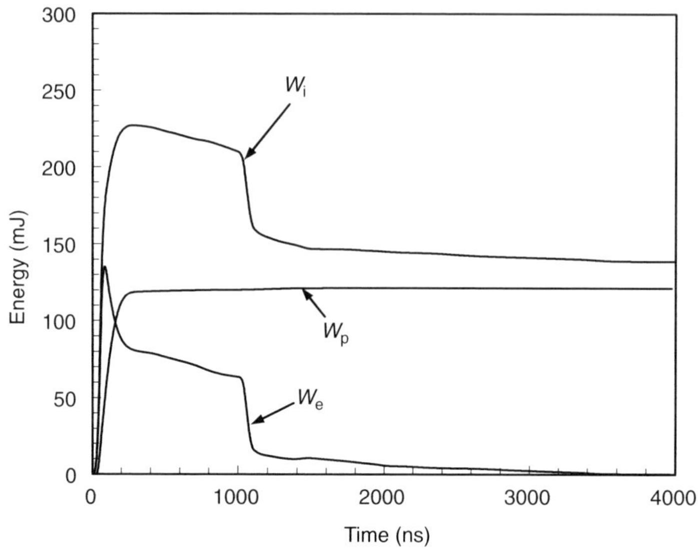

4.7 History of internal energy, elastically stored energy and plastically dissipated energy.

sipated energy sharply increases to 150 mJ and finally remains steady at around 180 mJ.

The internal energy includes the elastically stored energy (W_e) and plastically dissipated energy (W_p) (ABAQUS, 1998). Figure 4.7 shows how the total internal energy of different modes changes as the shock waves propagate through the material. The plastically dissipated energy dramatically increases to 110 mJ in a period of 200 ns and remains almost steady at 122 mJ after 2000 ns, while the elastically stored energy gradually reduces from 100 mJ to 30 mJ after 1000 ns. The saturation of the plastically dissipated energy implies that no more plastic deformation occurs in the material after a solution time of 2000 ns.

4.4.5 Steady dynamic stress state

Owing to dissipation and damping, the dynamic stress state will gradually achieve the converged solution in a certain solution period. However, the dynamic stresses are extremely slow to achieve the residual stresses because of high frequency local oscillation of the results around the final converged results. As mentioned in Chapter 3, in order to obtain the residual stress field in the specimen, the two finite element codes (ABAQUS/Explicit and ABAQUS/Standard) are integrated together to provide a most effective and efficient numerical solution. When the dynamic solution becomes

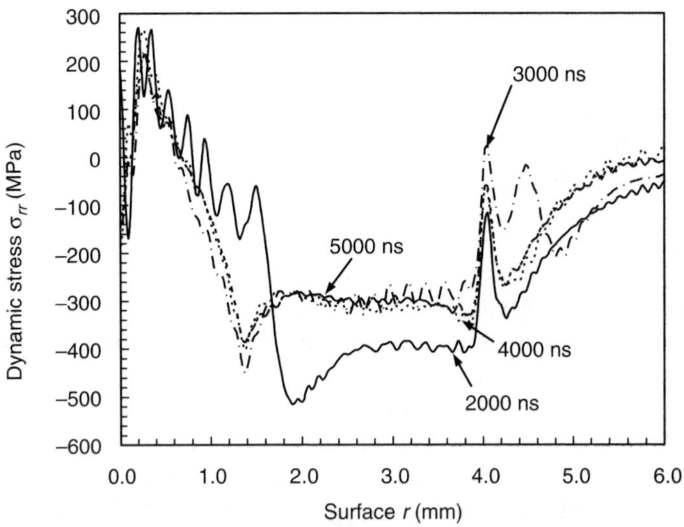

4.8 Surface dynamic stress (σ_{rr}) after different periods of solution time.

steady and no more plastic deformation occurs after a certain solution period from ABAQUS/Explicit, the deformed body containing the transient dynamic stress state is imported into the ABAQUS/Standard code. The static analysis is performed in ABAQUS/Standard to determine the residual stress field in static equilibrium. In the simulation of multiple LSP processes, the residual stress and strain states after the first LSP become the initial stress and strain states in the material for the second one, repeating this procedure for the third and so on. The residual stress is obtained for each of multiple LSP through the static equilibrium in the ABAQUS/Standard code when importing the transient dynamic stress and strain states from ABAQUS/Explicit to ABAQUS/Standard.

To allow plastic deformation to occur fully in the material, the solution time must take much longer than the duration of pressure pulse owing to the reflection and interaction of various shock waves propagating in the material. In the analysis, the solution was run for two orders of period longer than the pressure pulse duration to insure all plasticity has occurred in the material. Figure 4.8 shows the surface dynamic stress (σ_{rr}) profiles at the end of four periods of solution time. The dynamic stress profile is clearly diversified between the solution periods of 2000 ns and 3000 ns, but after the solution period of 4000 ns, the stress profile gradually becomes a steady one even though some 'ringing' occurs locally in the stress profile, approaching the converged results. Thus, based on all considerable factors mentioned above, the solution time was set at 4000 ns in the dynamic analysis.

4.4.6 Stress wave

As a round laser spot was used in the LSP process, the radial (σ_{rr}) and tangential dynamic stresses ($\sigma_{\theta\theta}$) in the three-dimensional space are axisymmetric. Figure 4.9(a) shows propagation of dynamic stresses (σ_{rr} and σ_{yy}) along the centreline ($r = 0$) at 200, 400 and 800 ns after the commencement of the pressure pulse. The maximum axial compressive stress at 200 ns is about 1750 MPa, but the radial compressive stress is only around 750 MPa. The magnitude of axial compressive stress is more than twice the radial compressive stress during shock wave propagation. In general, the compressive stresses in the both directions attenuate in magnitude with time as plastic deformation occurs within the material.

Figure 4.9(b) shows propagation of dynamic stresses (σ_{rr} and σ_{yy}) on the material surface at 200, 400 and 800 ns, respectively. Stress waves emitted from the perimeter of the impact spot gradually propagate to the centre. The peak magnitude of radial stresses (σ_{rr}) is about 1000 MPa, which is much higher than that of axial stresses (σ_{yy}) (slightly oscillating around the zero). After the stress waves are merged and reflected at the centre, they gradually attenuate in the both directions in magnitude with time as plastic deformation occurs within the material. After the shock waves have dispersed, the deformation and the radial compressive stresses remain. Compared with the axial stresses (σ_{yy}), the radial stress (σ_{rr}) is particularly important in LSP because it ultimately becomes the residual stress parallel to the treated surface.

4.4.7 Residual stress distribution

Figure 4.10 shows a 3D profile of predicted residual stresses in the radius and depth directions, impacted at a spot size of $r_p = 4$ mm. The distribution of predicted surface residual stress is depicted in Fig. 4.11. The experimentally measured compressive residual stresses from Ballard (1991) and the FE simulation by Braisted and Brockman (1999) for the 35CD4 steel, are also shown in Fig. 4.11. The experimental data indicate that the compressive residual stress is approximately zero at the centre of the spot and that there is a lack of compressive residual stress in the centre area of the treated zone. This phenomenon may be attributed to the simultaneous focusing of shock waves to the centre of impact zone, emitting from the edges of area under impact (Ballard, 1991; Masse and Barreau, 1995a). To avoid such a phenomenon, the shape of the impact spot used for an LSP process can be an unsymmetrical one, such as a square spot (Peyre *et al.*, 1996, Masse and Barreau, 1995a). The results of simulation, in Fig. 4.10 and 4.11, clearly show that a residual tensile stress zone obviously exists on the surface near the spot centre and a sharp oscillation of stress occurs at the spot edge. The stresses at both locations are sensitive to the density of finite elements. The

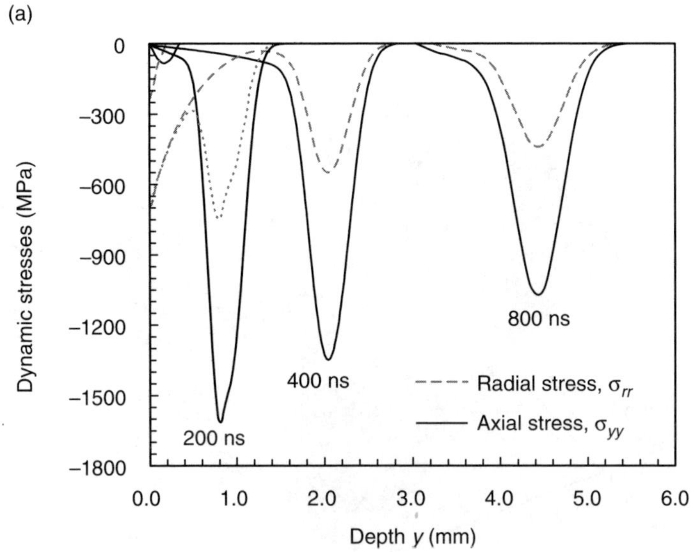

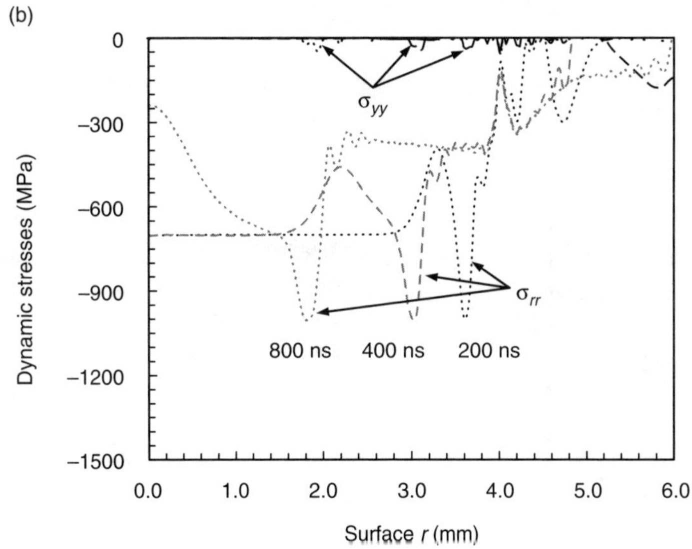

4.9 Dynamic stresses at 200, 400 and 800 ns. (a) In depth along centre line ($r = 0$), (b) on the surface.

higher magnitude of residual stress was observed when increasing the density of finite element mesh in these regions. Although the early simulation by Braisted and Brockman (1999) also reported similar behaviour, the mechanisms are not clearly understood at this stage.

The simulation results were also evaluated using two different pressure–time histories, a Gaussian pulse and a triangular ramp, respectively.

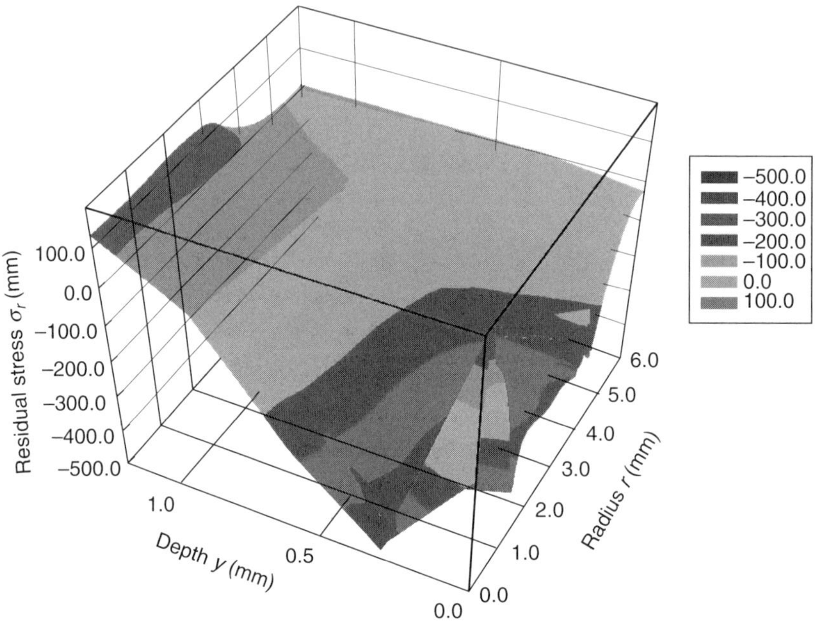

4.10 Distribution of residual stresses along the radius and in depth for single LSP.

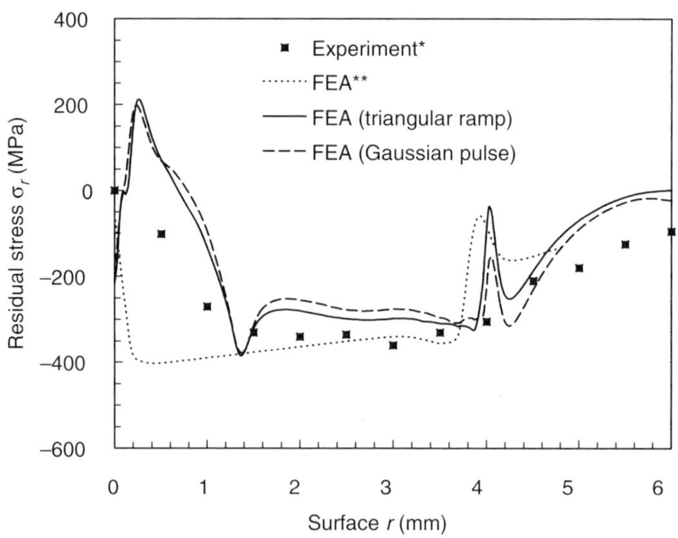

4.11 Surface residual stress profiles for single LSP. (*Source*: * Ballard, 1991; ** Braisted and Brockman, 1999).

Table 4.6 Peak surface residual stress and maximum plastically affected depth from experiments,* analytical* and FEM results

35CD4 30HRC steel	Experiment*	Analytical*	FEM
Surface residual stress, σ_r (MPa)	–360.0	–635.0	–390.0
Plastically affected depth, L_p (mm)	1.0	0.47	0.86

* Ballard, 1991.

The surface residual stress profiles from both pressure–time histories are almost identical to each other. In general, the predicted residual stresses at the rest part of surface correlate quite well with the experimental data.

Table 4.6 shows the peak surface residual stress from the experiment (Ballard, 1991) as well as those from analytical data (Ballard, 1991) and the present FEM simulation. The experimental data indicates a peak compressive residual stress of 360 MPa associated with a maximum plastically affected depth of 1 mm. The affected depth is the depth of the compressive stress field rather than the depth of plastic zone (or plasticised zone). The same terminology is used throughout the book when addressing the results from the experiments and simulations. The prediction shows that the peak compressive residual stress and the maximum plastically affected depth, L_p, (at $r = 1.4$ mm) are 390 MPa and 0.86 mm, a difference of 11% (higher) and 20% (smaller) from the experimental data, respectively. However, the analytical model produces a peak compressive residual stress of 80% higher and a plastically affected depth of 50% smaller than experiment.

The distribution of residual stress in the depth for single LSP from both the experiment (Ballard, 1991) and the simulation is shown in Fig. 4.12. The experimental data was approximately measured at $r = 3.5$ mm in the depth direction (y). The predicted residual stresses at $r = 3.5$ mm agree quite well with the experimental data, and most of experimental data fall between the two simulated residual stress profiles at $r = 1.4$ and 3.5 mm, respectively. The plastically affected depth below the impact surface for single LSP is shown in Fig. 4.13. The plastically affected depth along the radius is almost constant at 0.86 mm.

4.4.8 Parameter study

Multiple laser shock peening processes

Figure 4.14 shows the residual stress profiles for multiple LSP at the same spot. The peak compressive residual stress is increased to 460 MPa with the maximum plastically affected depth (L_{p2}) reaching 1.15 mm after two

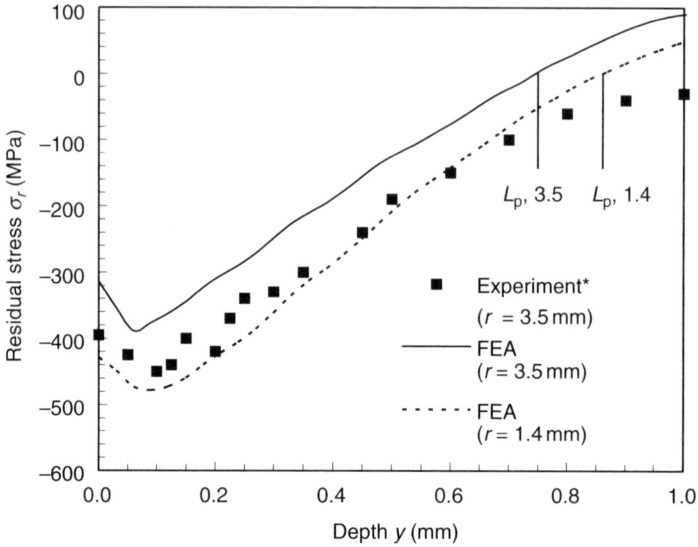

4.12 Distribution of residual stresses (r = 1.4 and 3.5 mm) in depth for single LSP (*Source*: * Ballard, 1991).

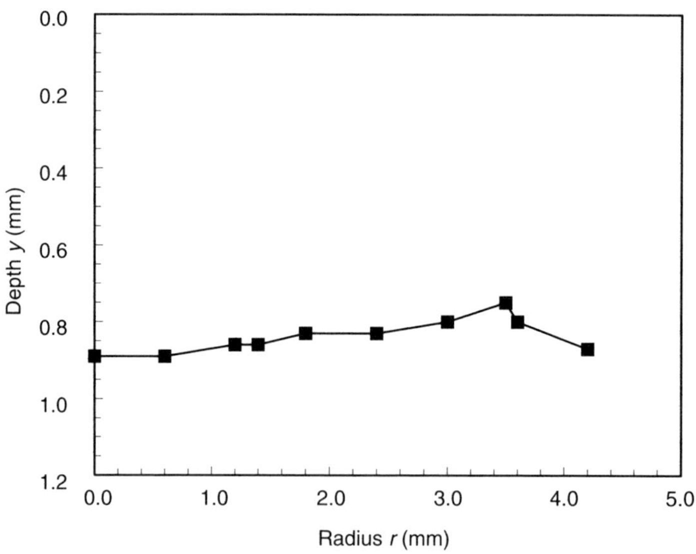

4.13 Distribution of maximum plastically affected depth (L_p) against radius.

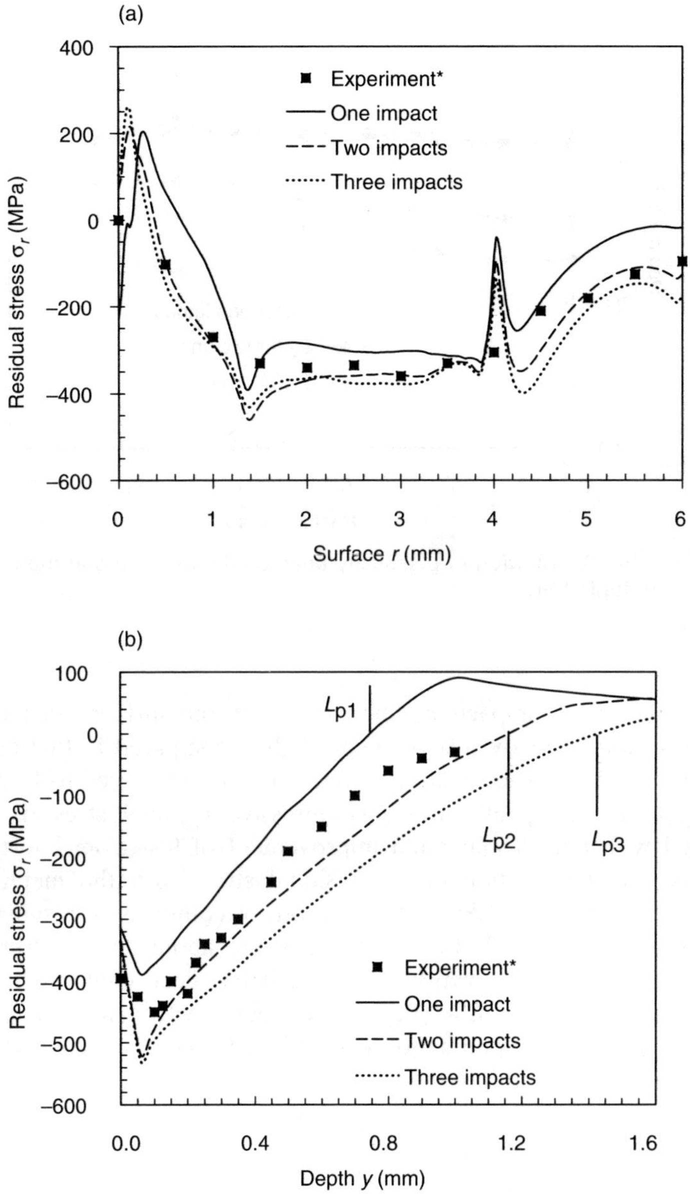

4.14 Distribution of residual stresses for multiple LSP. (a) On the surface, (b) in depth ($r = 3.5$ mm). (*Source*: * Ballard, 1991 for single LSP).

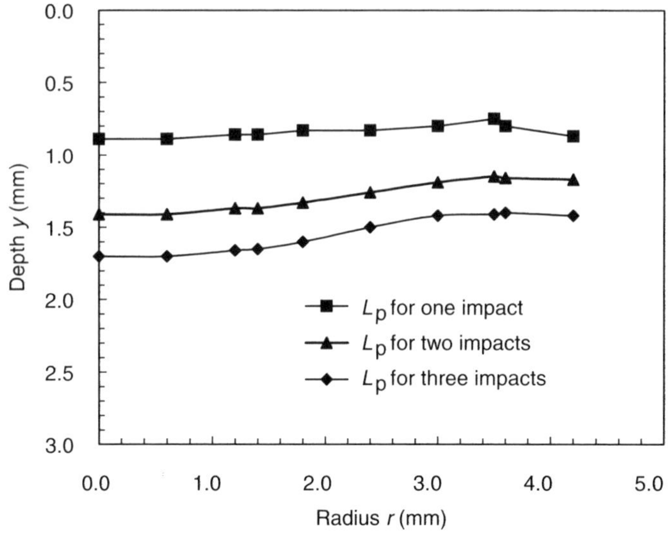

4.15 Distribution of plastically affected depth (L_p) along the radius for multiple LSP.

impacts. For three impacts on the same spot, the surface residual stress profile is almost steady, just changing slightly compared to that after two impacts, but the plastically affected depth (L_{p3}) is increased to 1.4 mm.

In general, the peak surface compressive residual stress is slightly changed with only a maximum improvement of 9% from one to three impacts. The penetration of the residual stress into the material was improved by multiple LSP with the depth of compressive residual stress increased by about 50% for two impacts and about 80% for three LSP impacts, respectively, shown in Fig. 4.15. It has been confirmed in many experimental studies of LSP (Peyre and Fabbro, 1995*b*; Fabbro *et al.*, 1998) that an increase in the number of repeated LSP impacts increases the depth of plastic deformation into the metallic materials.

Peak pressure

The plasma pressure pulse induced by LSP is a function of laser power density (Berthe *et al.*, 1997). The increase in the laser power density results in an increase in the magnitude of pressure pulse on the material surface (Fabbro *et al.*, 1990; Devaux *et al.*, 1993). To evaluate effects on the residual stress field with respect to changes in pressures, the peak pulse pressure used for the simulation was assumed to be 3.5 GPa according to an increase in the laser power density to 12 GW/cm^2, estimated from equation [3.5]. In

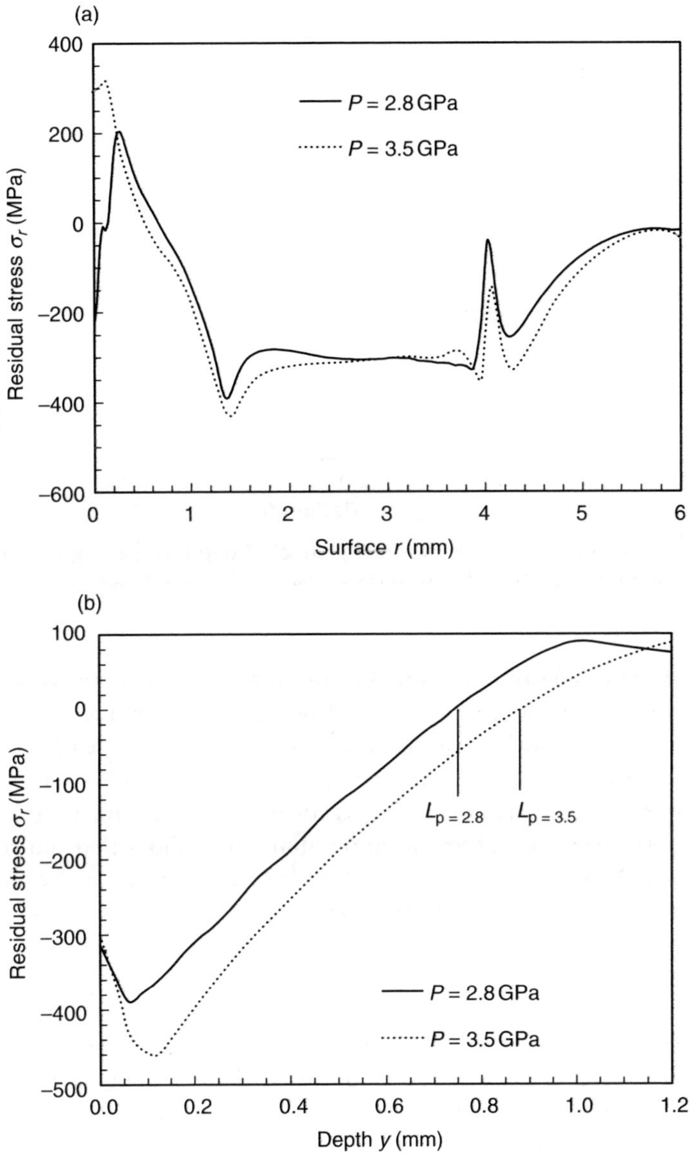

4.16 Distribution of residual stresses with respect to different peak pressures for single LSP.

addition, the simulation for each case was accomplished using the same pressure pulse duration of 100 ns for the pulse pressure–time history.

Figure 4.16(a) shows that the surface residual stress profiles are slightly changed with only an increase of 10% in the peak compressive stress as a

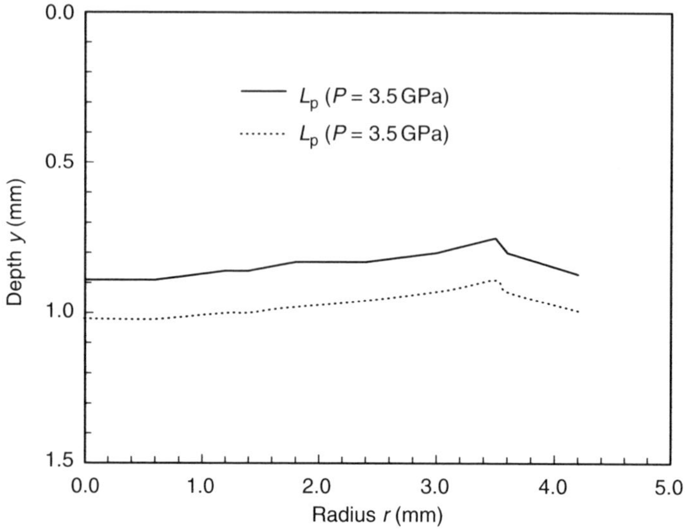

4.17 Distribution of plastically affected depth (L_p) along the radius with respect to different peak pressures for single LSP.

result of increasing the pressure, but the tensile residual stress is induced near the centre of the impact zone. However, the magnitude of the residual stress is obviously increased in the depth with an increase of almost 15% in the plastically affected depth, shown in Fig. 4.16(b) and Fig. 4.17, respectively. The predicted increase in the plastically affected depth correlates well with the experimental observation on aluminium alloys that the compressive residual stress could be driven deeper below the surface by increasing the pulse pressure (Clauer, 1996; Peyre *et al.*, 1996).

Pressure duration

As a laser system can deliver a wide range of pulse durations (between 0.1 and 50 ns) for LSP, the laser pulse duration directly controls the pressure pulse duration (Cottet and Boustie, 1989; Devaux *et al.*, 1993; Gerland and Hallouin, 1994; Couturier *et al.*, 1996). In order to evaluate the residual stress fields with respect to changes in the pressure duration, three periods of pressure duration with a peak pressure (2.8 GPa) were introduced in the simulation, with FWHM = 25, 50 and 100 ns, respectively.

Figure 4.18(a) shows that the surface compressive residual stress is significantly increased when using short pressure duration. The peak compressive residual stress for FWHM = 25 ns is about 594 MPa, which is 52% higher than that for FWHM = 50 ns and almost twice as high as that for FWHM = 100 ns. However, the plastically affected depth is significantly

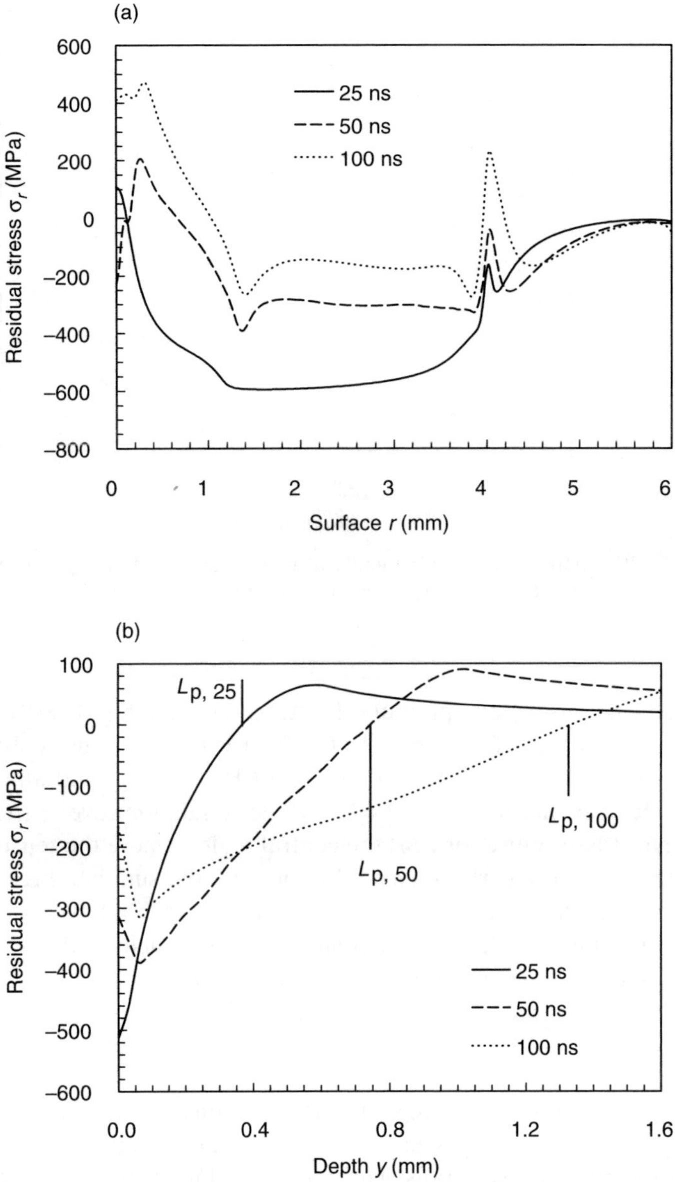

4.18 Distribution of residual stresses with respect to different pressure durations for single LSP. (a) On the surface, (b) in depth ($r = 3.5$ mm).

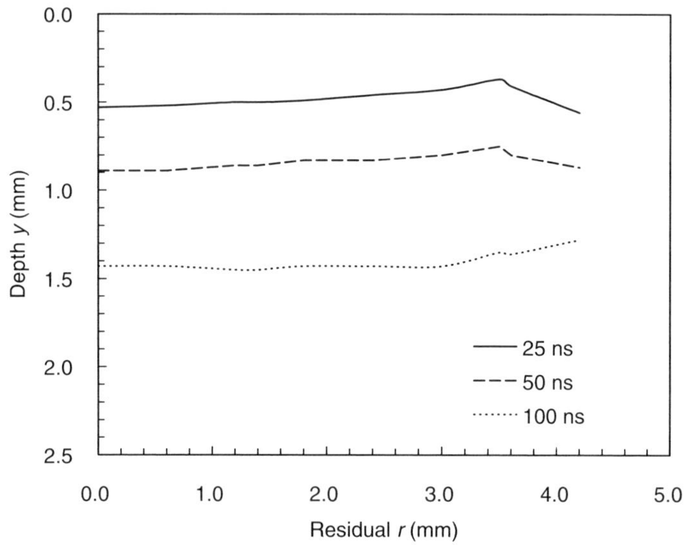

4.19 Distribution of plastically affected depth (L_p) along the radius with respect to different pressure duration for single LSP.

increased when using long pressure duration, shown in Fig. 4.18(b) and Fig. 4.19. The plastically affected depth (L_P) for FWHM = 100 ns is about 1.34 mm, which is 79% deeper than that for FWHM = 50 ns, and almost four times as deep as that for FWHM = 25 ns. The simulation reveals that, when using a short laser pulse for LSP, the central region where the tensile residual stress occurs is clearly reduced. In addition, the simulated results also agree with the experimental observation on a 12% chromium stainless steel that the maximum surface residual stress could be achieved with a short laser pulse (Fabbro *et al.*, 1998; Peyre *et al.*, 2000*a*).

Laser spot size

Using the same impact presure and LSP conditions, the simulation was performed to evaluate changes in residual stress fields caused by variations in the diameter of laser spots. The results in Fig. 4.20(a) show that the surface residual stress profiles change significantly when the spot size, r_p, is expanded from 1 to 4 mm. As a result, the maximum magnitude of surface compressive residual stress is clearly increased from 300 to 400 MPa. Meanwhile, the central region is affected by the diameter of spot with a small tensile residual stress for a large spot size.

Changing the laser spot size can also increase the depth of plastic deformation in the material. Figure 4.20(b) shows distributions of residual stress

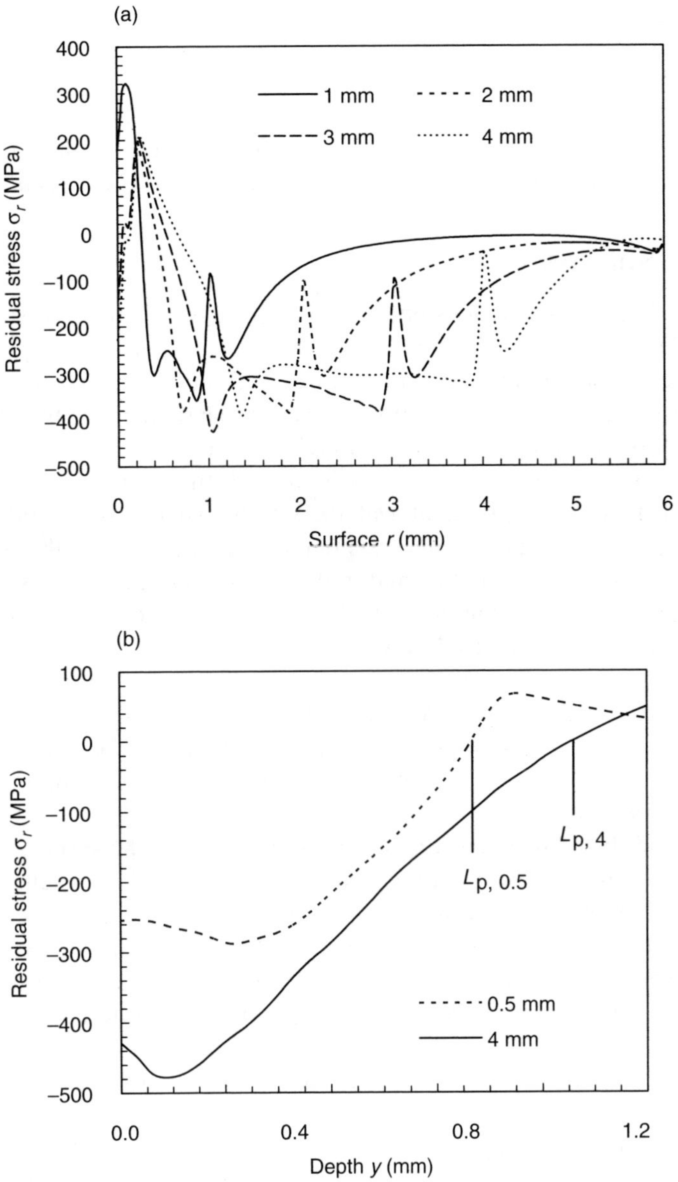

4.20 Distribution of residual stresses for laser spot of different radius in single LSP. (a) On the surface, (b) in depth.

in the depth direction from the surface point where the peak surface residual stress occurs for $r_p = 0.5$ and 4 mm, respectively. The plastically affected depth, L_p, is increased by 20% when increasing the size of laser spot from $r_p = 0.5$ to 4 mm. This phenomenon agrees with the experimental investigation on the 55C1 steel that plastically affected depth could be strongly reduced with a small impact configuration (0.5–1 mm) (Peyre *et al.*, 1998*b*).

4.5 Summary

In this chapter, two-dimensional finite element simulations of single and multiple LSP on a 35CD4 steel alloy have been evaluated using the simulation methodology discussed in Chapter 3. The effects of some key parameters in finite element simulations, such as mesh refinement, bulk viscosity and material damping, have been carefully evaluated and the suitable parameters for the analyses have been identified. In order to determine the residual stress field in the material, the steady dynamic stress state after a certain period of solution time was determined by tracking the evolution of energy dissipation in the material. The mechanical responses of the material for single and multiple LSP with respect to changes in LSP conditions are summarised as follows:

- The axial dynamic stress in the depth direction is much higher in magnitude than the radial one. The dynamic stresses in the both directions attenuate in magnitude with time as the plastic deformation occurs within the material.
- The predicted residual stresses in the depth or on the surface correlate quite well with experimental data. But the simulated results indicate that there is a small residual tensile stress zone on the surface near the center of impact spot and oscillation of stress occurs at the edge of the spot. The reason may be attributed to the stress wave focusing at the centre of the spot, which leads to a complicated interaction of stress waves. To avoid a lack of compressive residual stress in the central area of the treated zone, the shape of the impact spot can be changed to a square or unsymmetrical one.
- The simulated results indicate that the compressive residual stress can be increased and driven deeply below the surface by multiple LSP on the same spot.
- The compressive residual stress can also be driven deeply below the surface by increasing the pressure, but the surface compressive residual stress is not clearly increased as a result of increasing the peak pressure.
- The surface compressive residual stress and plastically affected depth are dependent on the pressure duration. The surface compressive residual stress can be increased using a short pulse duration for the process.

However, the increase in the plastically affected depths can only be achieved using a long pulse duration for the process.

- The plastically affected depth can be substantially reduced when using a small laser spot ($r_p < 1\,\text{mm}$). In addition, the peak compressive residual stress remains almost unchanged as a result of decreasing the laser spot size.

5 Three-dimensional simulation of single and multiple laser shock peening

5.1 Introduction

When a round laser spot is used for LSP, it may lead to a lack of compressive residual stress in the central area of the treated zone, as confirmed by experimental studies and simulation in the previous chapter. To avoid this drawback, a square laser spot can be used. The objective of this chapter is to evaluate the residual stress distribution in a metal alloy induced by single and multiple LSP with a square laser spot using the three-dimensional (3D) dynamic finite element method (FEM). The predicted residual stresses for single LSP are correlated with those from experiments and from an analytical model. The studies are focused on improving the understanding of dynamic stresses in the metal alloy during LSP as well as the effects of some essential factors on the residual stress field, such as laser spot size, pressure magnitude and duration, and number of LSP impacts.

5.2 Experimental

In the experiment of Ballard and his colleagues (Ballard *et al.*, 1991), the surface of a metal specimen (35CD4 50HRC steel) was irradiated by a square laser spot of 5×5 mm with laser power density of 8 GW/cm^2 and duration of 30 ns. The specimen surface was coated with black paint and the plasma was confined by a water overlay. The residual stress in the depth and on the surface of specimen was measured by means of X-ray diffraction, using successive electrolytic polishing.

For the purpose of evaluation, the material is assumed to be homogeneous, isotropic and elastic–perfectly plastic. Mechanical properties of material are summarised in Table 5.1. The pressure profile of induced plasma has a Gaussian temporal shape with a full width at half maximum (FWHM) of 50 ns and its peak pressure was 3.0 GPa (Ballard *et al.*, 1991). Owing to the narrow duration of the pressure pulse used in LSP, the pressure–time history can be explicitly simplified into a triangular ramp

Table 5.1 Mechanical properties of 35CD4 50HRC steel alloy (Ding and Ye, 2003b)

Properties	Value	Unit
Density (ρ)	7800	kg/m^3
Poisson's ratio (ν)	0.29	
Elastic modulus (E)	210	GPa
Lamé's constant (λ)	112.4	GPa
Lamé's constant (μ)	81.4	GPa
Hugoniot elastic limit (HEL)	2.1	GPa
Dynamic yielding strength (σ_y^{dyn})	1.24	GPa

with the pressure increasing linearly to a peak in the first 50 ns and then decaying linearly during the following 50 ns, as confirmed in Chapter 4.

5.3 Analytical model

As addressed in Chapters 3 and 4, an analytical model to predict the surface residual stress induced by LSP was developed by Ballard (1991). The model was assumed to be an elastic–plastic half space, with some basic assumptions including: (1) the shock-induced deformation is uniaxial and planar; (2) the pressure induced by the laser pulse is uniform in space; (3) the material obeys the Von Mises yielding criterion; and (4) work hardening and viscous effects are ignored. According to the model, the surface plastic strain, ε_p, can be written (Ballard, 1991) as:

$$\varepsilon_p = \frac{-2\text{HEL}}{3\lambda + 2\mu}\left(\frac{P}{\text{HEL}} - 1\right) \qquad [5.1]$$

where HEL is the Hugoniot elastic limit, P is the pressure and λ and μ are the Lamé's constants. In this case, substituting the values of HEL, P, λ and μ into equation [5.1], the surface plastic strain, ε_p, is 0.0036. To determine the residual stress field in the material, the plastically affected depth, L_p, for LSP is defined (Ballard, 1991) as:

$$L_p = \left(\frac{C_{el}C_{pl}\tau}{C_{el} - C_{pl}}\right)\left(\frac{P - \text{HEL}}{2\text{HEL}}\right) \qquad [5.2]$$

where C_{el} and C_{pl} are the elastic and plastic velocities and τ is the pressure pulse duration. In this case, C_{el} and C_{pl} calculated from equations [3.33] and [3.34] in Chapter 3 are 5940 and 4620 m/s, respectively. Substituting the values of HEL, P, τ (50 ns), C_{el} and C_{pl} into equation [5.2], the plastically affected depth, L_p, is 0.22 mm.

When using a square laser spot on the material, the surface residual stress can be defined by (Ballard, 1991):

$$\sigma_x^{\text{surf}} = \sigma_y^{\text{surf}} = \sigma_0 - [\mu\varepsilon_p(1+\nu)/(1-\nu)+\sigma_0]\left[1-\frac{4\sqrt{2}}{\pi}(1+\nu)\frac{L_p}{a}\right] \quad [5.3]$$

where a is the edge length of a square laser spot and σ_0 is the initial surface residual stress, set as zero in this study. Substituting L_p, ε_p, ν and μ into equation [5.3], the surface residual stress is –474 MPa.

5.4 Finite element model

A 3D dynamic finite element (FE) model was developed to simulate the process of a square laser spot impacting on the material surface. The simulation of short-duration shock wave propagation and the resulting residual stress in the material were accomplished using ABAQUS Explicit and Standard codes. As the model is symmetric, subjected to a symmetric uniform pulse pressure, a quarter of the configuration was used instead of the full one to perform the FE calculation, to assure computational efficiency.

Symmetric boundary conditions were employed on the x–0–z and y–0–z planes, respectively. The FE model, shown schematically in Fig. 5.1 with boundary conditions, consists of both finite and infinite elements and the infinite elements are assumed to be elastic elements used as non-reflecting boundaries. Eight-node finite elements can undergo non-linearity with large deformation to cope with high-pressure impact. The finite element domain is adjusted, if the residual stress field is close to the boundary between the finite and infinite element domains in a preliminary analysis.

Figure 5.1(b) shows the dimensions of the finite element area and the impact zone in the x–0–z plane. In the x-direction, x_f (= 5 mm) is defined as the edge of the finite element area, while x_p (= 2.5 mm) is defined as the edge of the impact zone, which is a half of the square laser spot edge. In addition, in the y- and z-directions, y_f and z_f are set to be equal to x_f, while y_p is equal to x_p as a result of the square laser spot.

To evaluate the effect of mesh refinement on the simulation results, three models with coarse to fine meshes were applied for the analyses, defined as Ma, Mb and Mc, respectively. Their configurations are described in Table 5.2. The mesh density is defined as a ratio of the element length, L^e, to the impact zone size, x_p. Ma is a coarse mesh having a mesh density of 10% with total elements of 9200, being much coarser than Mb and Mc, shown in Table 5.2.

Similar analysis procedures for the single and multiple LSP processes used in the two-dimensional (2D) simulation described in Chapter 4 are applied. The pulse pressure induced by the laser shot is assumed to be

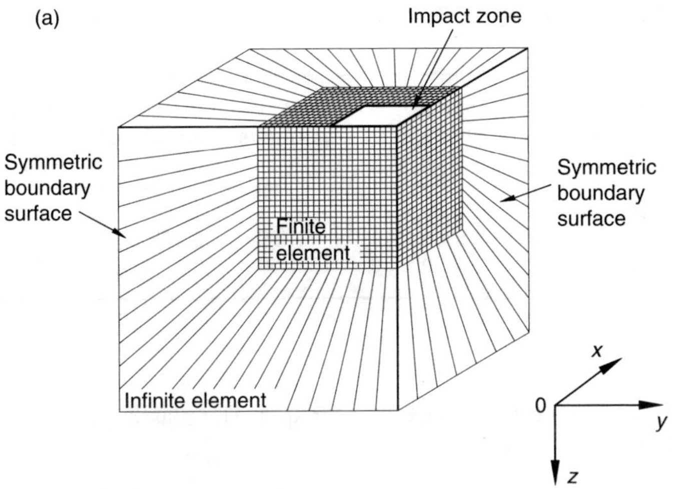

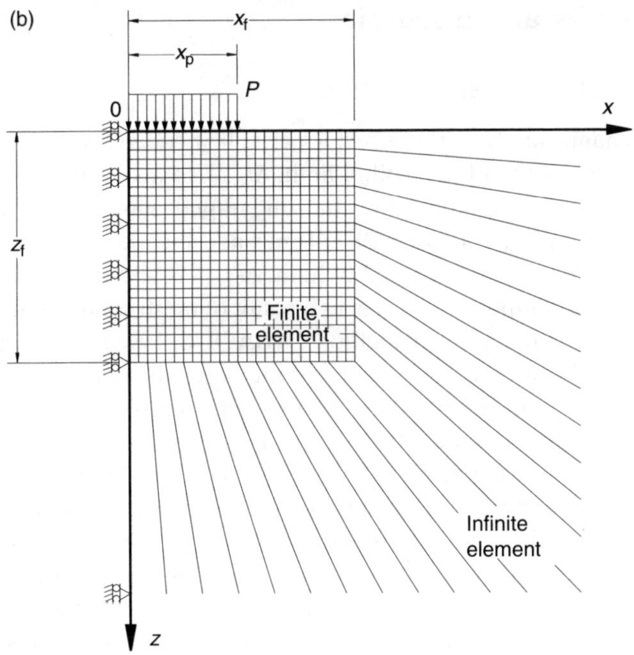

5.1 Three-dimensional finite element model for simulation of LSP. (a) Boundary conditions, (b) model geometry in the plane (*x*–0–*z*) (Ding and Ye, 2003b).

Table 5.2 Configurations of FE models of different mesh densities (Ding and Ye, 2003b)

FE model	Finite element	Infinite element	Total elements	Element length (L^e) (mm)	Mesh density (L^e/x_p) (%)
Ma	20 × 20 × 20	3 × 20 × 20	9200	0.25	10 (coarse)
Mb	40 × 40 × 40	3 × 40 × 40	68800	0.125	5 (moderate)
Mc	50 × 50 × 50	3 × 50 × 50	132500	0.1	4 (fine)

uniform over the entire surface of laser spot. The plastic yielding is defined by the Von Mises criterion.

5.5 Results and discussion

5.5.1 Effect of mesh refinement

During a dynamic analysis, the ABAQUS/Explicit code sets the time increment, Δt, to be less than the stability limit, Δt_{stable}, to avoid numerical instability. For instance, for model Ma, the undamped elastic wave speed of material, $C_d = (E/\rho)^{1/2}$, is about 5.19×10^3 m/s and the element length, L^e, is 0.25 mm. Thus, the estimated stability limit, Δt_{stable}, is about 4.8×10^{-8} s.

The estimated computational cost as a result of mesh refinement is rather straightforward in the explicit method (ABAQUS, 1998). As discussed in Chapter 4 for 2D cases, mesh refinement increases the computational cost by increasing the number of elements and reducing the smallest element dimension (i.e. stability limit). If the mesh is refined by a factor of 2 in all three directions, the computational cost increases by a factor of 8 as a result of the increase in number of elements and further by a factor of 2 as a result of the decrease in the time increment. In the analyses, if Ma is chosen as a benchmark, its time increment and central processing unit (CPU) time are defined as S and C, respectively. After the mesh is refined, the time increment is decreased to $0.5S$ for Mb and $0.4S$ for Mc and the corresponding CPU time is increased to $16C$ for Mb and $39C$ for Mc, respectively.

The dynamic stresses (σ_{xx}) resulting from Ma, Mb and Mc, respectively, along the x- and z-axes, respectively, after a solution period of 4000 ns, are shown in Fig. 5.2. The peak surface compressive stress is approximately 320 MPa for Ma, 370 MPa for Mb and 341 MPa for Mc. The stress profiles for Mb and Mc are similar even though there are more local oscillations in the surface stress profile for Mc than for Mb. Moreover, the in-depth stress profile resulting from Ma is clearly different from those of Mb and Mc. In

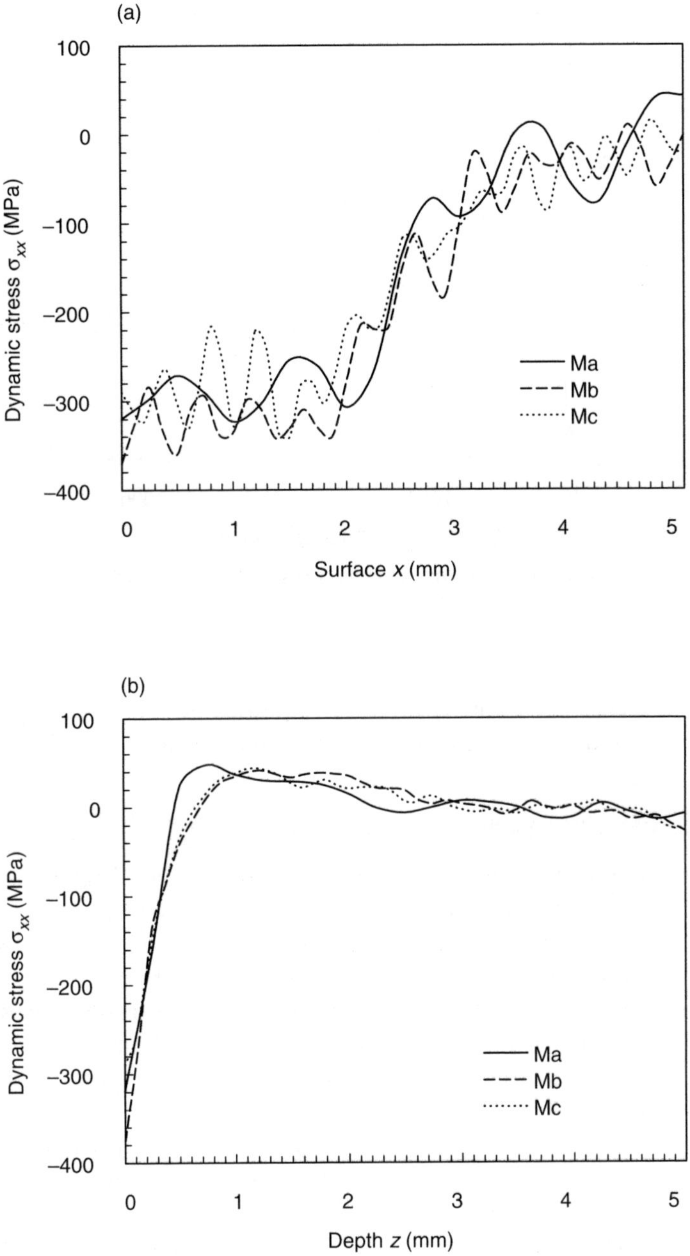

5.2 Dynamic stress distribution from different FE models. (a) σ_{xx} on the surface, (b) σ_{xx} in depth.

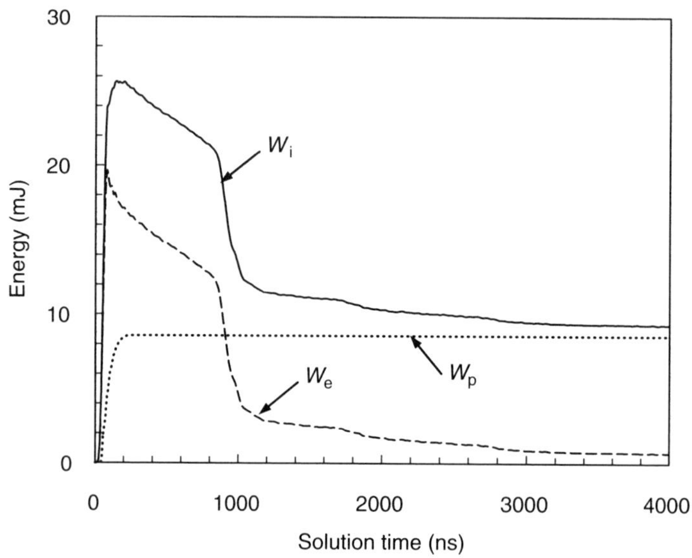

5.3 History of internal energy, elastically stored energy and plastically dissipated energy.

order to achieve reasonably accurate solutions but ensure computational efficiency, model Mb is selected for the further evaluation.

5.5.2 Steady dynamic stress state

Figure 5.3 shows how the total internal energy (W_i), consisting of elastically stored energy (W_e) and plastically dissipated energy (W_p), changes as the shock waves propagate through the material. The plastically dissipated energy dramatically increases to 8.4 mJ in the initial period of 200 ns and remains steady at 8.6 mJ after 2500 ns, while the elastically stored energy reduces to about 1 mJ after 1000 ns. The saturation of plastically dissipated energy implies that no further plastic deformation takes place in the material after 2500 ns.

The surface dynamic stress (σ_{xx}) along the x-axis, varying with the solution time, is depicted in Fig. 5.4. The dynamic stress profile changes quite clearly in the solution period from 1000 to 2000 ns, but after a solution time of 4000 ns the stress profile becomes almost steady, approaching the converged results. Therefore, it is reasonable to set 4000 ns as the solution time for the dynamic analyses in order to ensure computational efficiency, when the transient stress and strain states are imported from ABAQUS/Explicit to ABAQUS/Standard to calculate the residual stress field in static equilibrium.

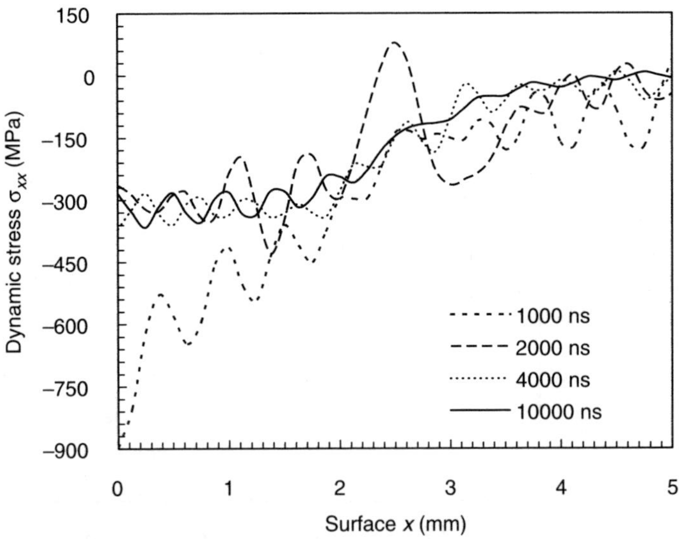

5.4 Surface dynamic stress (σ_{xx}) profiles after different periods of solution time.

5.5.3 Dynamic stress

As a square laser spot was used for the LSP process, σ_{xx} along the x-axis is equivalent to σ_{yy} along the y-axis and vice versa. Figure 5.5(a) shows the dynamic stresses, σ_{xx} or σ_{yy}, along the depth (z) of specimen ($x, y = 0$) at the end of solution periods of 200, 400 and 800 ns, respectively. The magnitude of stress at 200 ns is around 850 MPa, which is about 12% higher than that at 400 ns and 26% higher than that at 800 ns. The attenuation of stress in magnitude with time in the z-direction is attributed to plastic deformation within the material.

The dynamic stresses, σ_{xx} and σ_{yy}, along the x-axis on the surface ($z = 0$) at the end of 200 and 400 ns are depicted in Fig. 5.5(b). The magnitude of σ_{xx} at 200 ns is about 823 MPa, which is about 28% higher than that of σ_{yy}, while σ_{xx} attenuates in magnitude by 24% at 400 ns. The compressive stresses, σ_{xx} and σ_{yy}, are the dominant dynamic stresses which become eventually the compressive residual stresses, σ_x and σ_y, parallel to the treated surface.

5.5.4 Residual stress

Table 5.3 summarises the peak surface residual stress and the maximum plastically affected depth from experiment (Ballard, 1991) as well as those simulated by the analytical model (Ballard, 1991) and the 3D simulation.

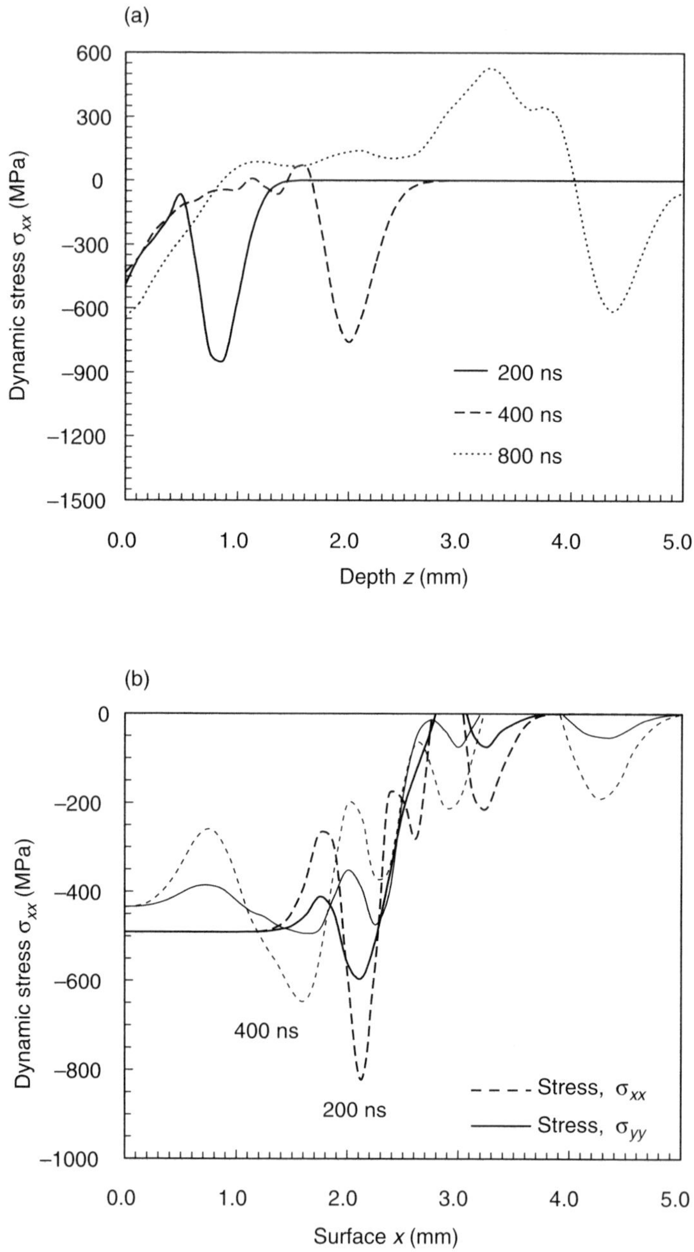

5.5 Dynamic stresses (σ_{xx} or σ_{yy}) profiles in the process. (a) In depth (z) along the centre line (x, $y = 0$), (b) along the x-axis.

Table 5.3 Surface residual stress from experiment, analytical and 3D FE simulation

35CD4 50HRC Steel	Experiment*	Analytical*	FEA (Mb)
σ_x (MPa)	–355.0	–474.0	–331.0
L_p (mm)	0.80	0.22	0.62

* Ballard *et al.*, 1991.

The experimental data show that the peak compressive residual stress (σ_x) is about 355 MPa with a maximum plastically affected depth of 0.80 mm. The 3D simulation predicts a peak compressive residual stress (σ_x) of about 331 MPa with a maximum plastically affected depth of 0.62 mm, being 7% and 23% lower than the experimental measurements, respectively. For the results from the analytical model (Ballard, 1991), the peak residual stress and plastically affected depth are 34% higher and 70% lower than the experimental ones, respectively.

Figure 5.6(a) shows the surface residual stresses (σ_x and σ_y) along the x-axis from the experiments (Ballard, 1991) and the 3D simulation for single LSP. The predicted surface residual stress correlates quite well with the experimental data. The simulated results reveal that there is no lack of compressive residual stress at the centre of treated zone, unlike the LSP process using a circular laser spot evaluated in Chapter 4. Therefore, a square laser spot can lead to better results for the LSP process, avoiding the negative effect caused by the focusing of shock stress waves at the centre of the treated zone as a result of using a circular laser spot (Masse and Barreau, 1995*a*; Braisted and Brockman, 1999).

The in-depth residual stresses (σ_x and σ_y) from experiment (Ballard, 1991) and the 3D simulation for single LSP are plotted in Fig. 5.6(b). The predicted plastically affected depth (L_p) is about 0.62 mm, shown in Fig. 5.7, being 23% less than the measurement (0.80 mm). However, good correlation in distribution of in-depth residual stresses (σ_x or σ_y) can be observed between experimental data and the 3D simulation.

5.5.5 Multiple laser shock peening process

The surface and in-depth residual stress profiles resulting from multiple LSP at the same spot are shown in Figs. 5.8 and Fig. 5.9. The peak surface compressive residual stress (σ_x or σ_y) for single LSP is 331 MPa and the plastically affected depth is 0.62 mm. The peak surface compressive residual stress and plastically affected depth are significantly increased by 24% to 410 MPa and 37% to 0.85 mm, respectively, as a result of two impacts on

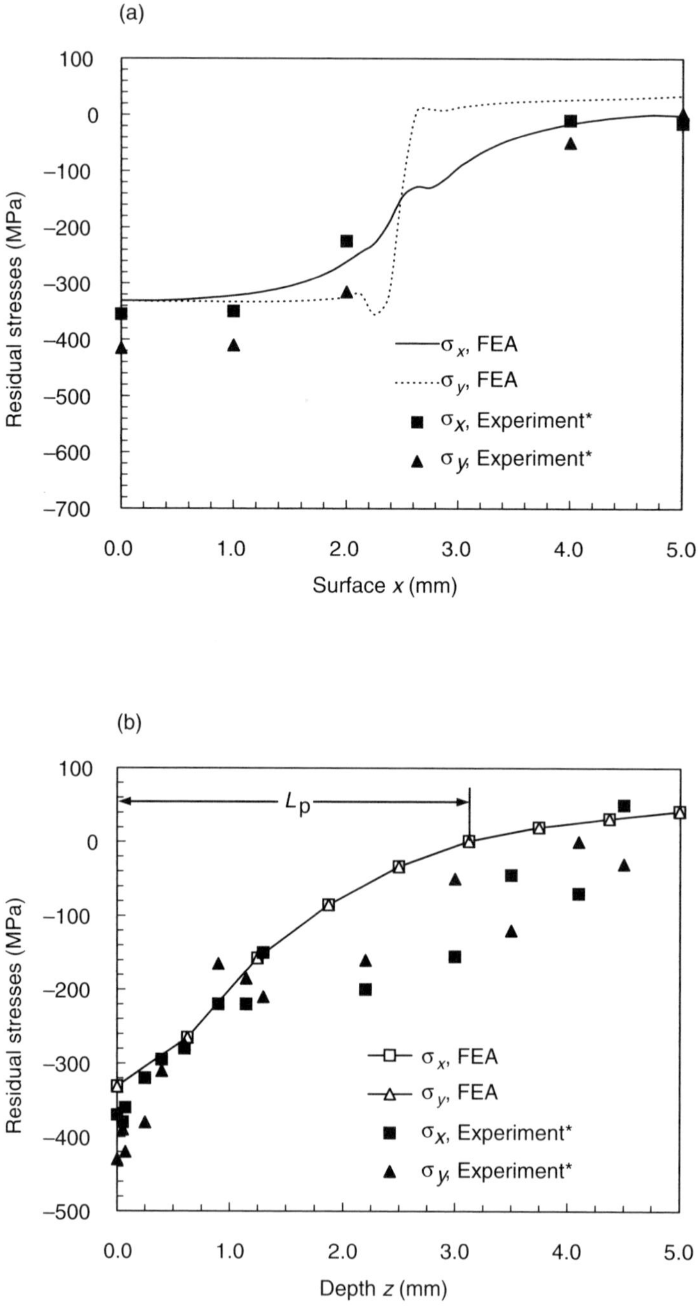

5.6 Residual stresses (σ_x and σ_y) after single LSP. (a) On the surface along the *x*-axis, (b) in depth along the *z*-axis. (*Source*: *Ballard *et al.*, 1991).

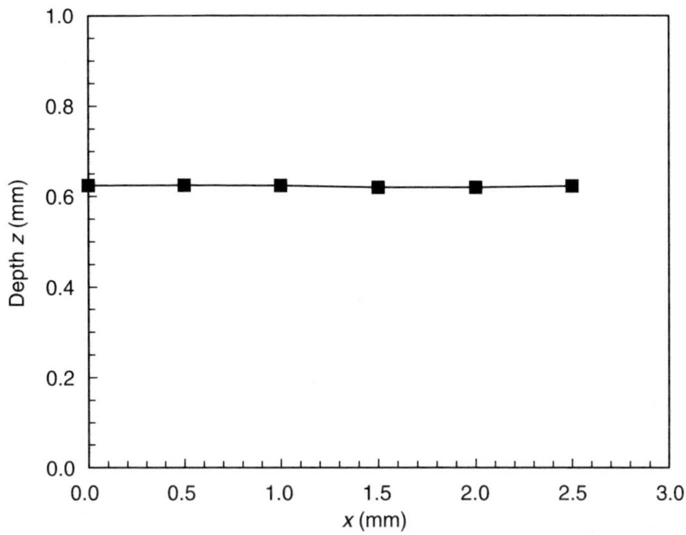

5.7 Profile of plastically affected depth (L_p) in the *x*-direction.

the same spot. After three impacts, the peak surface compressive residual stress is further increased up to 436 MPa and the plastically affected depth reaches 0.95 mm. The plastically affected depth increases almost linearly with the number of impacts on the same spot, while the increase in surface residual stress gradually reaches its saturation, correlating well with the observation from experiment (Peyre *et al.*, 1998b; Clauer *et al.*, 1996; Fabbro *et al.*, 1990).

5.5.6 Parameter study

In order to evaluate the residual stress field with respect to changes in impact pressures, the peak pressure used for the simulation was assumed to be 3.0, 4.0 and 5.0 GPa, respectively, in relation to the increase in laser power density. In addition, the simulation for each case was accomplished using the same duration of 100 ns for the pressure–time history.

Figure 5.10 shows the surface and in-depth residual stress profiles with respect to the increase in the impact pressure. Simulation with a peak pressure of 4.0 GPa produces the highest compressive residual stress of 655 MPa, which is 50% and 11% higher than that with a peak pressure of 3.0 GPa and 5.0 GPa, respectively. A saturated level of surface residual stress is found at around 600 MPa when the peak pressure is increased from 4.0 to 5.0 GPa. The saturation in the surface residual stress can be attributed to the magnitude of yield strength of the material (Peyre *et al.*, 1998a; Fabbro *et al.*, 1998). It has been observed in experimental studies that a peak

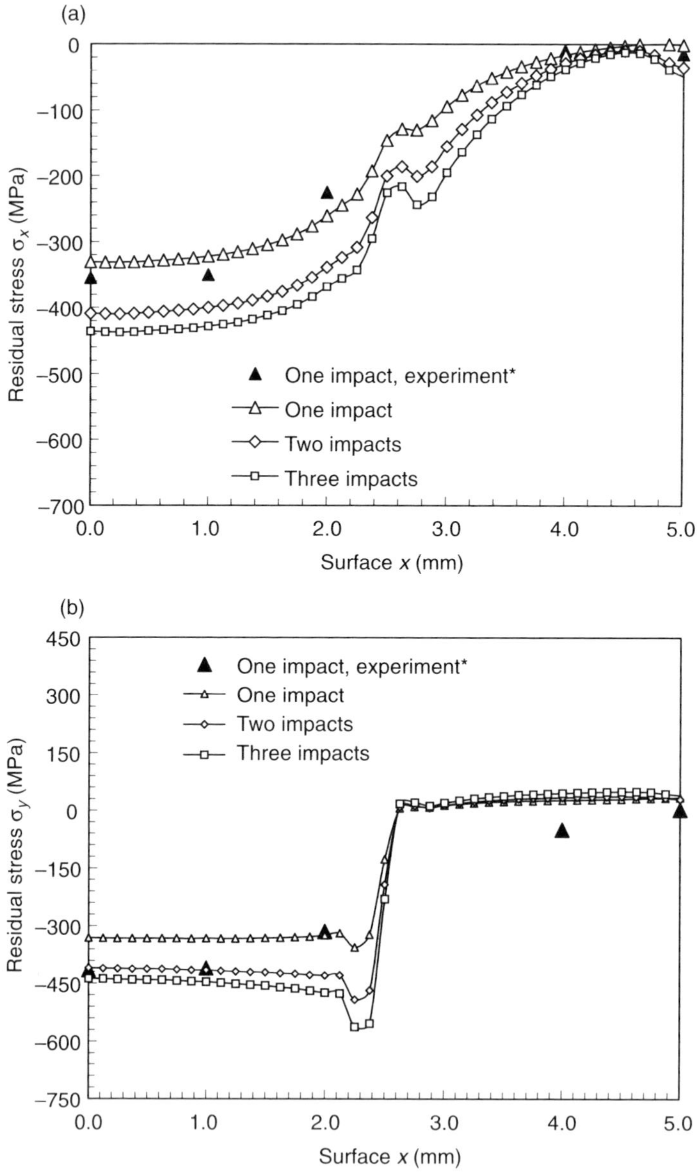

5.8 Surface residual stresses along the *x*-axis after multiple LSP. (a) σ_x, (b) σ_y. (*Source*: *Ballard *et al.*, 1991).

pressure normally in the range between 2 × HEL and 2.5 × HEL produces the best results for an LSP process (Peyre *et al.*, 1996; Fabbro *et al.*, 1998).

The plastically affected depth (Fig. 5.11) is about 0.62 mm for a peak pressure of 3.0 GPa. As a result of increasing the peak pressure from 3.0

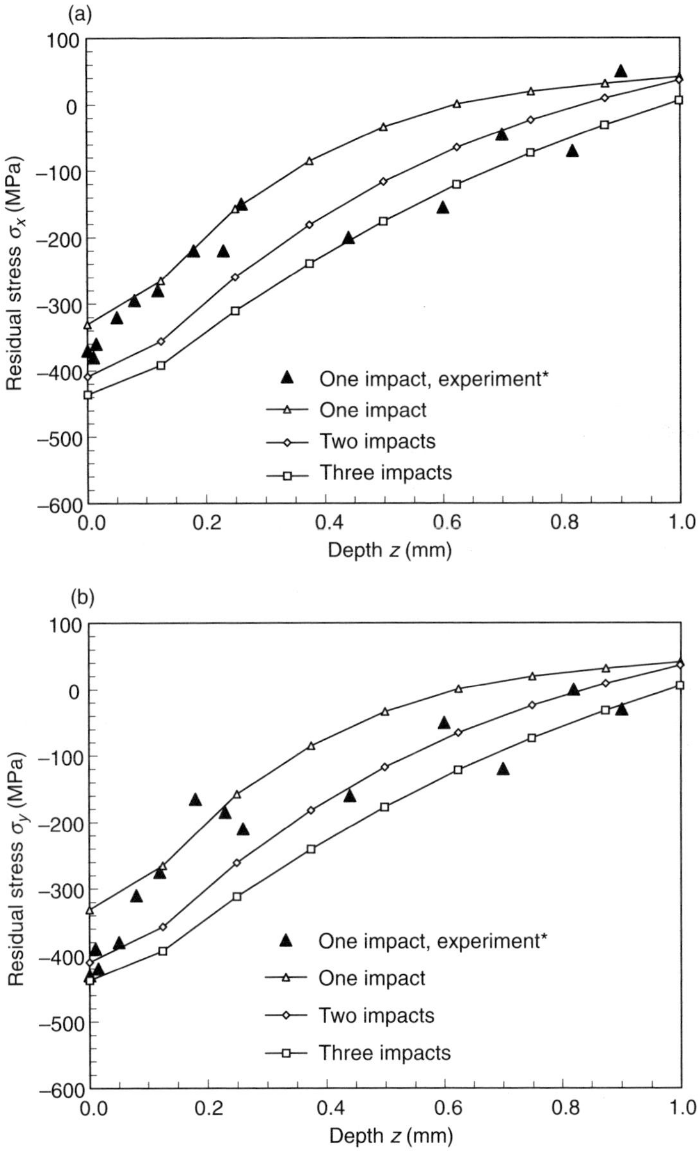

5.9 Residual stresses along the centre line (*z*-axis) after multiple LSP. (a) σ_x, (b) σ_y. (*Source*: *Ballard *et al.*, 1991).

to 5.0 GPa, the plastically affected depth is clearly increased by 24% to 0.81 mm for a peak pressure of 4.0 GPa and by 38% to 1.0 mm for the peak pressure of 5.0 GPa. The increase in the plastically affected depth in the material is in a good agreement with the experimental observation in aluminium alloys that the compressive residual stress could be driven deeper

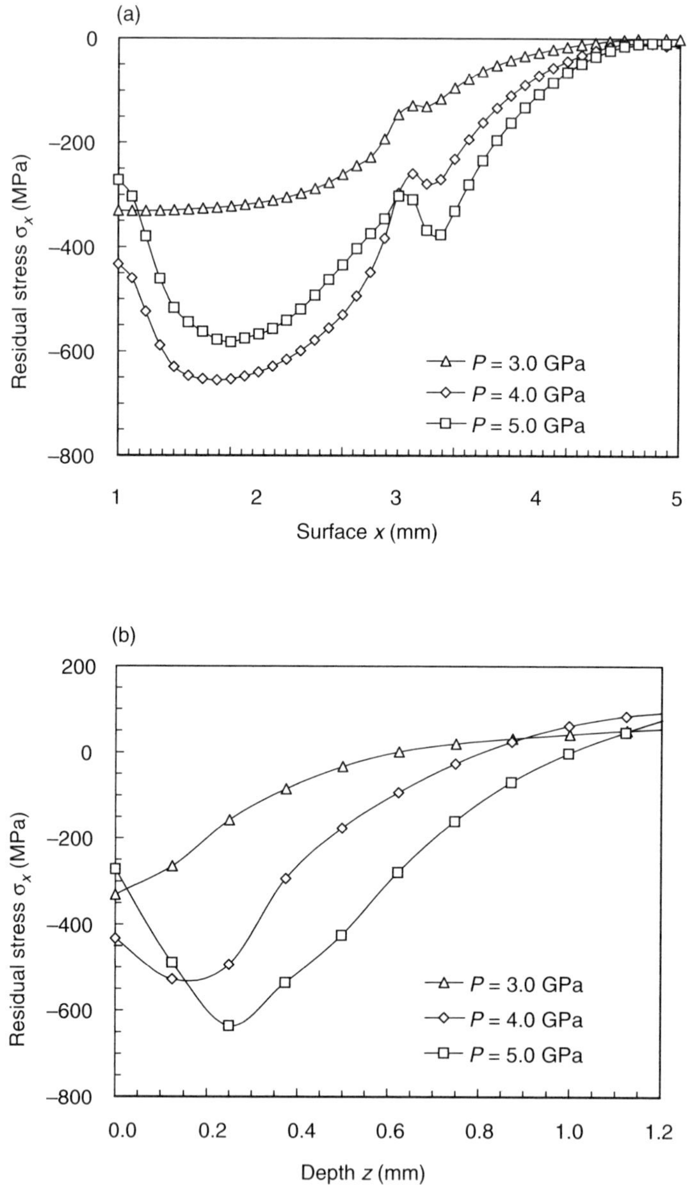

5.10 Distribution of residual stresses (σ_x) for different peak pressures after single LSP. (a) On the surface, (b) in depth.

below the surface by increasing the impact pressure (Montross *et al.*, 2002; Peyre *et al.*, 1996; Fabbro *et al.*, 1998).

To evaluate the residual stress field with respect to changes in the pressure pulse duration, four periods of pressure pulse duration with a peak

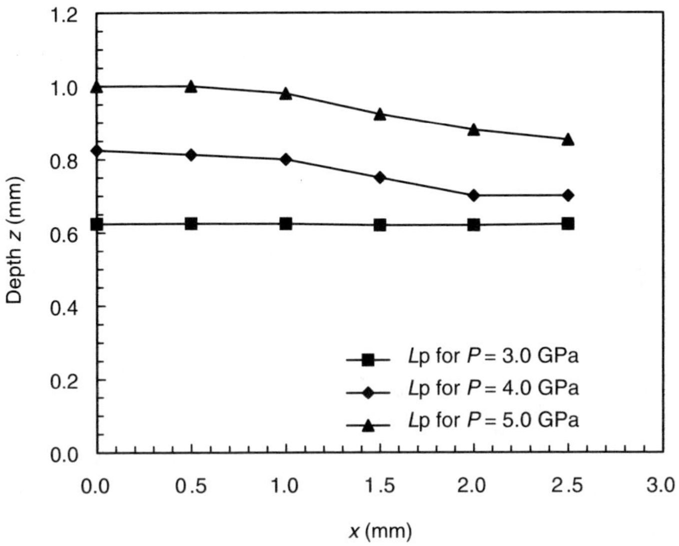

5.11 Profiles of plastically affected depths (L_p) in the *x*-direction for different peak pressures.

pressure of 3.0 GPa were introduced in the simulation, namely Sa, Sb, Sc and Sd, and the pressure duration at FWHM was defined as 10, 15, 25 and 50 ns, respectively.

Figure 5.12 shows distributions of surface residual stress along the *x*-axis and in-depth residual stress with respect to the different pressure duration. A duration of 50 ns (FWHM) produces a maximum surface compressive residual stress of 350 MPa, being 5% higher than that for 25 ns and 84% higher than that for 15 ns. In particular, there is no residual stress generated on the material surface when using a pressure duration of 10 ns (FWHM). The plastically affected depth is also greatest when a long pressure duration is used. With a duration of 50 ns (FWHM), the plastically affected depth is about 0.62 mm, which is 56% deeper than that for 25 ns and 64% deeper than that for 15 ns. The simulation reveals that the plastically affected depth is reduced when using short pressure duration, consistent with experimental observation (Peyre *et al.*, 1998b).

Simulation was also performed to evaluate changes in the residual stress field corresponding to variations in the laser spot when the peak pressure is 3.0 GPa and other LSP conditions are kept unchanged. The results in Fig. 5.13 (a) show that the surface residual stress profile changes significantly when the spot size, x_p, increases in the range 1 to 4 mm. The peak compressive residual stress is clearly increased by 14% from 290 to 330 MPa as a result of increasing x_p from 1 to 4 mm. However, it is clearly shown that the peak residual stress level remains almost steady at about 330 MPa when

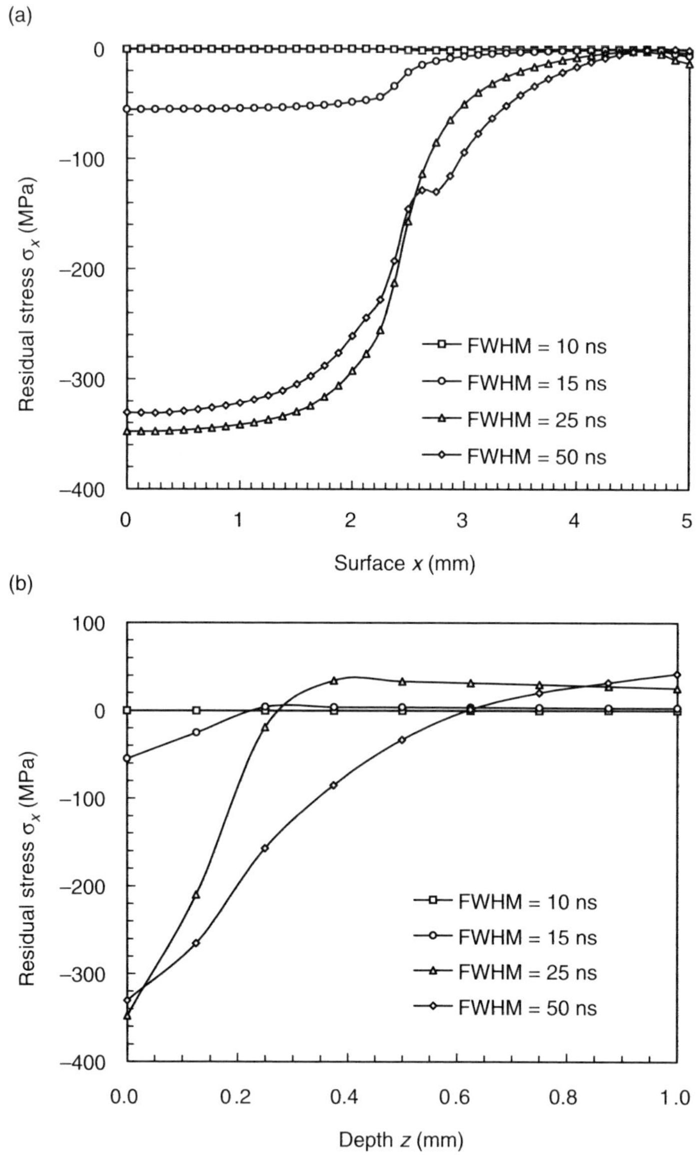

5.12 Distribution of residual stress (σ_x) for single LSP of different duration of pressure. (a) On the surface, (b) in depth.

x_p is increased from 2.5 to 4 mm. The distribution of in-depth residual stress (σ_x) in the material with respect to changes in laser spot size is plotted in Fig. 5.13 (b). The plastically affected depth (at $x = y = 0$) is almost steady at 0.62 mm, regardless of laser spot size.

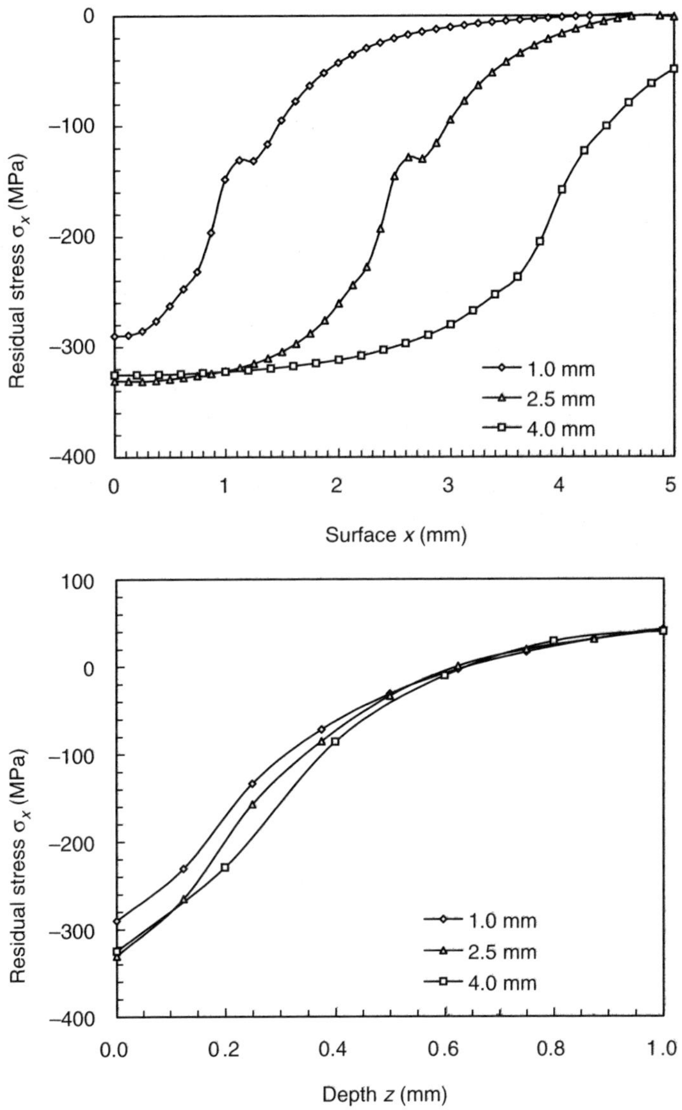

5.13 Residual stress (σ_x) profiles for single LSP of different laser spot sizes. (a) On the surface, (b) in depth.

5.6 Summary

In this chapter, 3D dynamic FE analyses were applied to study the distribution of residual stress in a metal alloy of 35CD4 50HRC steel after LSP treatment with a square laser spot. The simulated results for single LSP correlated reasonably well with the available experimental data. The simulation also revealed that using a square laser spot for LSP produces a better result in terms of distribution of surface residual stress than using a circular laser spot, which led to a lack of compressive residual stress at the centre of the treated zone, as revealed in Chapter 4.

It can be concluded from the simulation that the compressive residual stress is increased and driven more deeply below the surface with multiple LSP on the same spot. The plastically affected depth is also increased when using a high impact pressure for LSP, while a high impact pressure does not necessarily lead to a high magnitude of compressive residual stress. In addition, both the magnitudes of compressive residual stress and plastically affected depth are significantly dependent on pressure duration. The plastically affected depth is increased with long pressure duration, but almost zero surface residual stress is obtained when the pressure duration is very short (FWHM = 10 ns). In addition, the plastically affected depth is almost unchanged as a result of changing the laser spot size from 1 to 4 mm, while the residual stress increases slightly.

6 Two-dimensional simulation of two-sided laser shock peening on thin sections

6.1 Introduction

For a thick alloy component, LSP can be performed individually or simultaneously on any part of a surface. However, for a thin section, a laser pulse must be split, impacting simultaneously on opposite sides of the section and balancing impact forces. If the section subjected to one-sided LSP is thin enough, the peened spot can create a dimple on the irradiated side and a bulge on the opposite side. It can also cause spalling and fracture if the shock waves are strong enough. Meanwhile, if the laser pulse impacts on a large area, significant curvature or other distortion can be induced in the thin section (Clauer and Lahrman, 2001).

In military aircraft and spacecraft there are a large number of thin metal components that can be treated using two-sided LSP. The aim in this chapter is to provide a better understanding of LSP mechanisms in thin metal sections undergoing two-sided LSP using dynamic finite element (FE) simulations. The distribution of residual stress on the surface and in the interior of thin sections is simulated and mechanically affected critical areas through the thin sections are carefully evaluated. The results are correlated with the experimental data available in literature. The studies are also focused on obtaining a better understanding of two-sided LSP in relation to effects of some essential factors on the residual stress field, such as laser spot size, pressure, material geometry and number of LSP impacts.

6.2 Laser shock peening model

Clauer and Lahrman (2001) investigated the compressive residual stress in a 1.0 mm thick section of Ti-6Al-4V subjected to LSP on both sides simultaneously. The experiments were conducted under three processing conditions: low (one shot with laser power density of 5.5 GW/cm^2), intermediate (three shots with laser power density of 5.5 GW/cm^2) and high (three shots with 10 GW/cm^2). In the experiments, the high-energy laser pulse was split

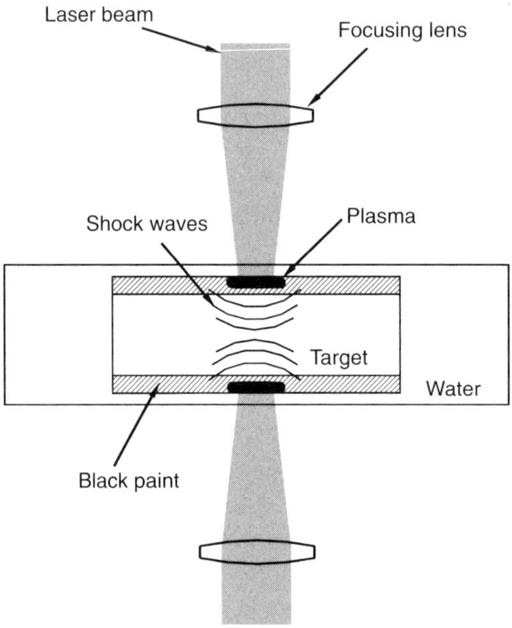

6.1 Schematics of laser shock peening on a thin section.

Table 6.1 Mechanical properties of Ti-6Al-4V alloy (Braisted and Brockman, 1999)

Density (ρ) (kg/m^3)	4500
Poisson's ratio (ν)	0.342
Elastic modulus (*E*) (GPa)	110
Hugoniot elastic limit (HEL) (GPa)	2.8
Dynamic yielding strength (σ_y^{dyn}) (GPa)	1.345

into two, which were simultaneously focused onto two sides of the thin section. The schematics of two-sided LSP are illustrated in Fig. 6.1. The mechanical properties of Ti-64Al-4V alloy are summarised in Table 6.1.

As discussed in Chapter 3, the laser-induced plasma generates a high pulse pressure because of the recoil momentum of the ablated material (Montross *et al.*, 2002; Fabbro *et al.*, 1998). In the confined mode with an overlay, assuming a constant absorbed laser power density, I_0, the peak pressure, P, is given (Berthe *et al.*, 1997; Fabbro *et al.*, 1998) by:

$$P(\text{GPa}) = 0.01\sqrt{\frac{\alpha}{2\alpha+3}}\sqrt{Z(\text{g/cm}^2\text{s})}\sqrt{I_0(\text{GW/cm}^2)} \quad [6.1]$$

where Z is the reduced acoustic impedance [$2/Z = 1/Z_1 + 1/Z_2$] of the confining (Z_1) and target (Z_2) materials and α is the efficiency of the plasma-material interaction. In the water confinement mode, the acoustic impedance (Z_1) of water is 1.65×10^5 g/cm^2 s, while the acoustic impedance (Z_2) of the target (Ti-6Al-4V) is about 2.75×10^6 g/cm^2 s. α is determined by calibrating the equation with the experimental data, varying over a range of 0.2 to 0.5 and is dependent on the overlay and other processing conditions (Clauer and Lahrman, 2001).

As the laser power density used for LSP was 5.5 GW/cm^2, the estimated peak pressure ranges from 3.5 to 5.0 GPa, assuming $\alpha = 0.2$ to 0.5. Since no other information for the LSP condition is available in literature, the laser spot size on each side of section was assumed to be in a range of 4–8 mm in diameter.

6.3 Finite element model

During an LSP process, mechanisms of stress waves while propagating through the thickness of a thin section are too complex to be evaluated by means of experimental instrument or analytical modelling. However, it is feasible to establish a two-dimensional (2D) FE model to simulate and investigate these complicated mechanisms in the section. The pressure–time history for the FEM simulation can be explicitly simplified into a triangular ramp, in which the pressure rises linearly to the peak in first 50 ns and then decays linearly during the following 50 ns, as addressed in Chapters 4 and 5. In the simulation, the pressure applied over the laser spot is assumed to be uniform. The metallic section is assumed to be elastic–perfectly plastic, with isotropic and homogeneous characteristics. The plastic yielding is assumed to follow the Von Mises criterion.

It is assumed that the thin section is large enough in its in-plane dimensions to be modelled as a semi-infinite solid of axisymmetry if a round laser spot is applied. The FE model in Fig. 6.2 consists of two kinds of elements, four-node finite elements and infinite elements. The infinite elements are assumed to be only elastic in behaviour and are used as non-reflecting boundaries. The finite elements can undergo non-linearity with large deformation to cope with high-pressure impact. As the geometry and loading profile in the model are symmetric for two-sided LSP, a quarter of the model was selected for the FE analysis to assure computational efficiency.

It is necessary to extend or adjust the region of the FE mesh if the calculated residual stress areas are close to the boundary between finite and infinite elements, as addressed in Chapter 4. The radius (L) of the FE mesh area was finally set to be twice the laser spot radius (r_0) and its depth (D) was the half of the section thickness. In order to evaluate the mesh sensitivity of FEA results, three FE models with coarse to fine mesh were

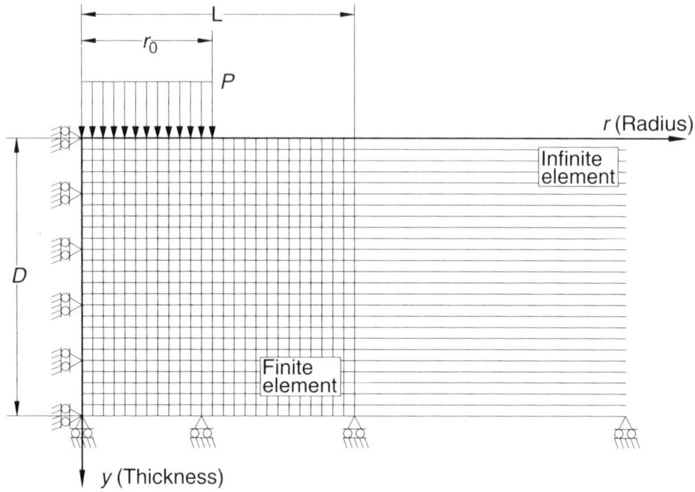

6.2 Schematics of an FE model for a thin section with axisymmetric boundary conditions (Ding and Ye, 2003a).

Table 6.2 Configurations of FE models (r_0 = 3 mm) of different mesh density (Ding and Ye, 2003a)

FE model	Finite element	Infinite element	Element size (L^e) (mm)	Mesh density (L^e/r_0) (%)
Ma	25 × 300	25 × 1	0.02	0.67 (coarse)
Mb	50 × 600	50 × 1	0.01	0.33 (intermediate)
Mc	75 × 900	75 × 1	0.0067	0.22 (fine)

employed in the simulations. Their configurations are depicted in Table 6.2. Mesh density is defined as the ratio of the element size (L^e) to the radius of the impact zone (r_0). If using a laser spot size of 6 mm (or r_0 = 3 mm) for the process, the mesh density for the coarse (Ma), intermediate (Mb) and fine (Mc) models is 0.67%, 0.33% and 0.22%, respectively. The same approaches as described in Chapter 4 for simulating single and multiple LSP were applied in the evaluation using the ABAQUS/Explicit and ABAQUS/Standard codes.

6.4 Evaluation of modelling

The sensitivity of simulation on FE mesh is evaluated in terms of surface dynamic stress, σ_{rr}, from the three models, i.e. coarse mesh (Ma) with 25 ×

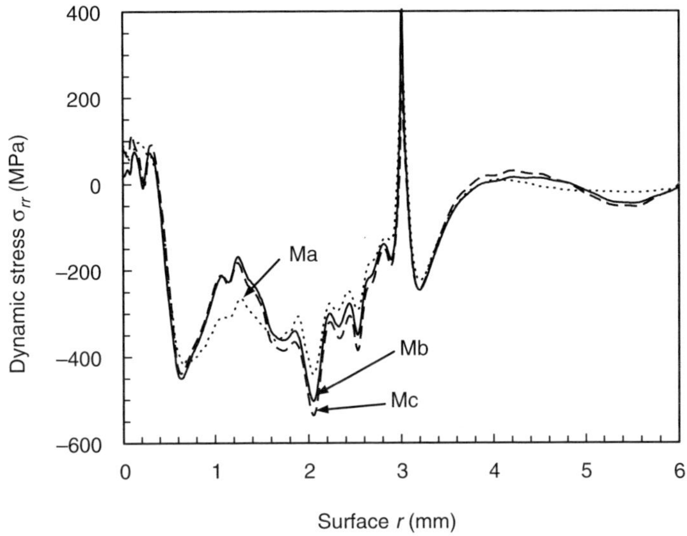

6.3 Dynamic surface stress profiles (σ_{rr}) from coarse (Ma), intermediate (Mb) and fine (Mc) meshed FE models.

300 finite elements, intermediate mesh (Mb) with 50 × 600 finite elements and fine mesh (Mc) with 75 × 900 finite elements in the finite element area. Figure 6.3 shows that the surface dynamic stress (after a solution period of 16 000 ns) from Mc is quite similar to that from Mb, but differs quite clearly from that from Ma. However, the surface dynamic stress profile almost approaches a consistent solution if the mesh density is further increased. In order to ensure computational efficiency, mesh Mb is selected for the further evaluation.

Figure 6.4 shows that the total internal energy of different modes dissipates as the shock wave propagates through the section. In the model, the elastically stored energy is around 400 mJ at the commencement of the LSP process and then decreases dramatically to 40 mJ after 5000 ns, while the plastically dissipated energy increases significantly in the initial period of 200 ns and remains steady at about 360 mJ after 15 000 ns. This saturation implies that no more plastic deformation occurs in the section after this period. Figure 6.5 shows that the surface dynamic stress profile (σ_{rr}) varies with the solution time. The dynamic stress profiles change quite clearly between the solution periods of 8000 ns and 16 000 ns, but after the solution time of 40 000 ns the stress profile becomes steady, approaching the converged results. Thus, the solution time is set as 40 000 ns in the dynamic analyses, in order to import the transient stress and strain states from ABAQUS/Explicit to ABAQUS/Standard to calculate the residual stress field in static equilibrium.

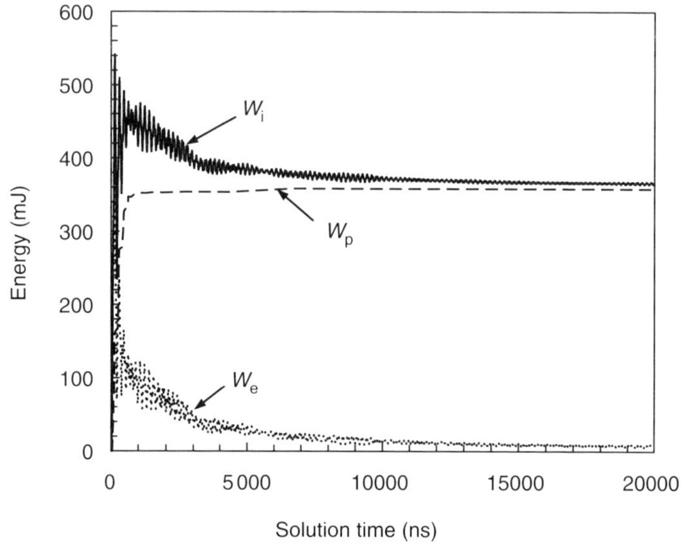

6.4 Evolution of internal energy, elastically stored energy and plastically dissipated energy.

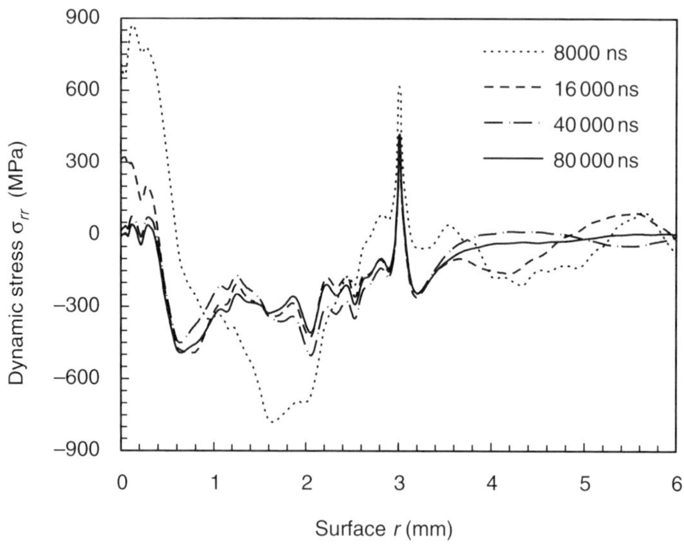

6.5 Surface dynamic stress (σ_{rr}) profiles at the end of four periods of solution time (8000, 16000, 40000 and 80000 ns).

As the experimental assessment of dynamic stress states within a thin section induced by two-sided LSP is very difficult to perform, the investigation of complicated shock wave interaction within the thin section is another interesting study. The simulated dynamic stress associated with

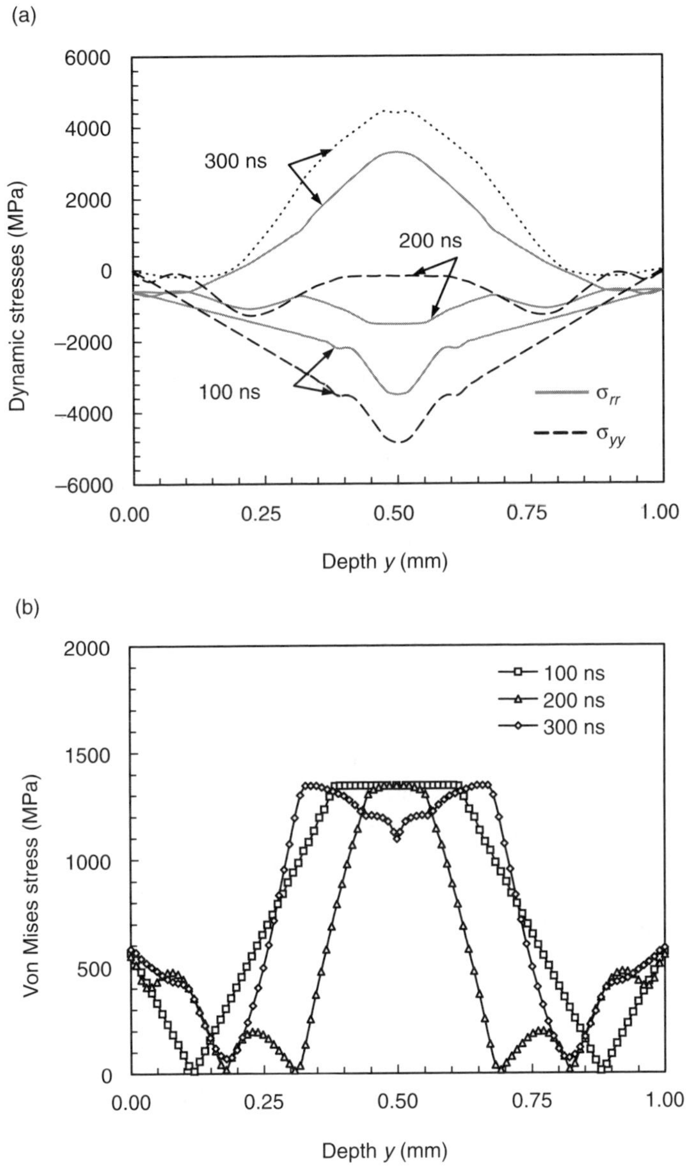

6.6 Dynamic stresses at 100, 200 and 300 ns along the centre line ($r = 0$) through the thickness. (a) Normal stresses, (b) Von Mises stress.

stress wave propagation through the section thickness along the centre line ($r = 0$) is illustrated in Fig. 6.6(a), for radial stress (σ_{rr}) and axial stress (σ_{yy}), captured at three instant periods of 100, 200 and 300 ns. At 100 ns, the compressive stress reaches the mid-plane ($y = 0.5$ mm), where the axial compressive stress is about 40% higher in magnitude than the radial one. When the stress waves are reflected back to the surface after 200 ns, the magnitudes of both axial and radial compressive stresses are significantly reduced in the mid-plane. However, when stress waves are reflected to the mid-plane again after 300 ns, high tensile stresses occur at the mid-plane and the magnitudes of both axial and radial stresses seem to be the same as those after 100 ns. High tensile dynamic stresses in the middle plane can be very harmful for the material because they may cause cracking or delamination at the location.

When dynamic stresses in the section reach the dynamic yield strength of the material, plastic yielding occurs at the location. The distributions of the Von Mises stress along the centreline ($r = 0$) at three instant periods of 100 ns, 200 ns and 300 ns are depicted in Fig. 6.6(b). At 100 ns, the peak Von Mises stress occurs in the mid-plane of the section. After this duration, unsaturated plastic deformation at various locations through the thickness increases with time before the stress waves attenuate to a level at which the plastic deformation becomes steady.

6.5 Effects of parameters

6.5.1 Laser spot size

The residual stress profile is dependent on LSP conditions, such as laser power density, laser spot size and laser pulse duration (Clauer *et al.*, 2001; Peyre *et al.*, 1998b). In order to evaluate changes in residual stress fields with respect to variations in the diameter of laser spot, three laser spot sizes in a range of r_0 = 2–4 mm were applied for the simulations. Figure 6.7(a) shows that the surface residual stress profiles change remarkably with an increase in laser spot size. However, the peak compressive stress level is almost constant at 450 MPa when the laser spot radius is increased from 2 to 4 mm.

Figure 6.7(b) shows the residual stress in the depth of the section when the laser spot size is increased from r_0 = 2 to 4 mm. The plastically affected depth (L_1 for r_0 = 2 mm, L_2 for r_0 = 3 mm and L_3 for r_0 = 4 mm) associated with the compressive residual stress varies slightly in a range of 0.1–0.14 mm because of the changes in laser spot size, although it can be argued that the plastically affected depth cannot be easily defined in this case.

6.5.2 Impact pressure and multiple laser shock peening

As discussed in previous chapters, increasing the laser power density also increases plasma pressure. Optimisation of plasma pressure on material

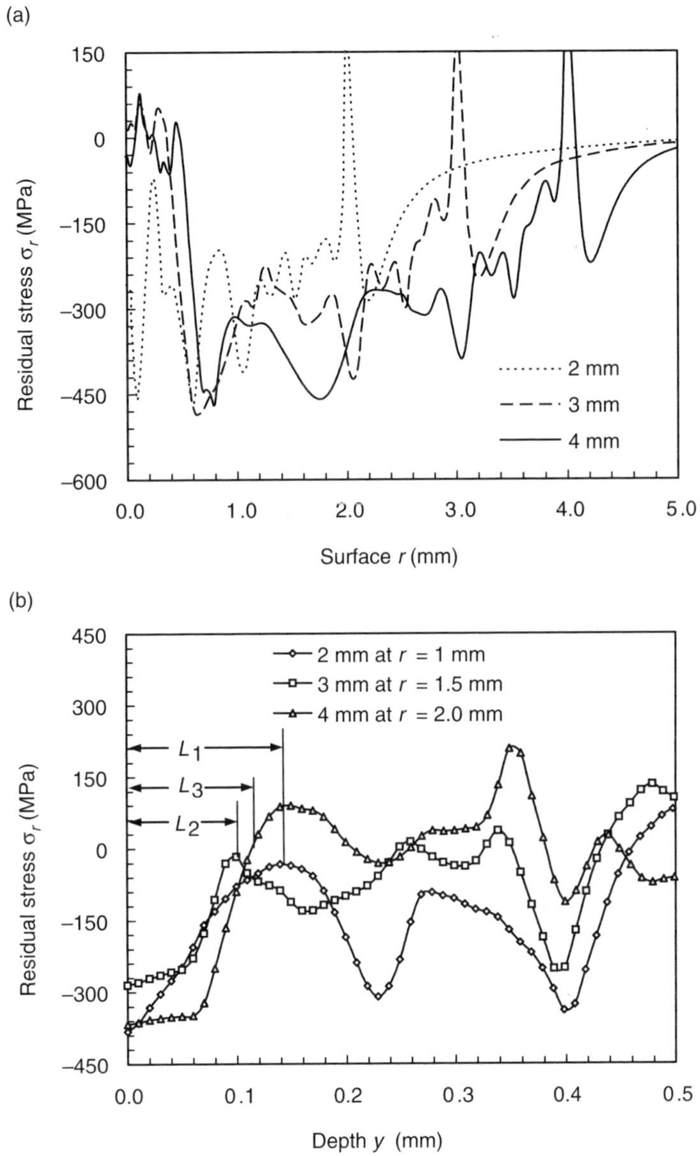

6.7 Residual stress distribution in the section for laser spot of different sizes. (a) On the surface, (b) in depth.

surface can play an important role in achieving the best treatment results (Masse and Barreau, 1995*a*; Montross *et al.*, 2002; Fabbro *et al.*, 1998). Figure 6.8 shows the residual stress profiles in the depth of a section (at r = 1.5 mm) corresponding to three different peak pressures, 3.5, 4.0 and 5.0 GPa, when the pulse duration of 100 ns and laser spot size of $r_0 = 3$ mm

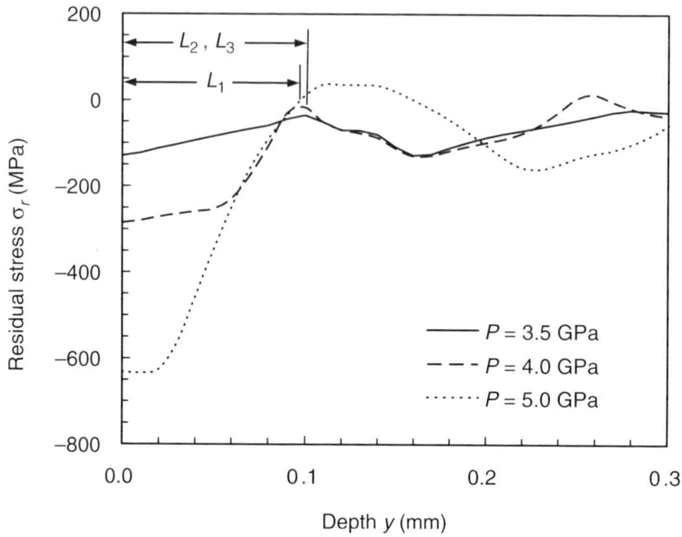

6.8 Residual stresses at $r = 1.5$ mm in the depth of section for different peak pressures.

were kept the same. The magnitude of surface residual stress is significantly increased when the peak pressure level increases. However, the plastically affected depth (L_1 for $P = 3.5$ GPa, L_2 for $P = 4.0$ GPa, L_3 for $P = 5.0$ GPa) associated with the compressive residual stress slightly changes in a range of 0.085–0.1 mm in the depth (at $r = 1.5$ mm), depicted in Fig. 6.8.

Figure 6.9 shows the distribution of compressive residual stress (σ_r) through the thickness in the section from both the simulation and the experiments (Clauer and Lahrman, 2001). The predicted in-depth residual stress profile is at $r = 1.5$ mm when the laser spot size and peak pressure used for the simulations were set as $r_0 = 3$ mm and $P = 4.0$ GPa, respectively. There is reasonable correlation between the experimental data and the simulated results. However, the plastically affected depth associated with the compressive residual stress is difficult to define properly from the simulation. A tensile residual stress zone with a maximum magnitude of about 140 MPa occurs in the region near the mid-plane. Clauer (1996) reported this phenomenon for LSP on a 1.5 mm thick 4340 steel sheet and it was found that the tensile residual stress at the mid-plane of the section was higher after five shots. The effects in this region are attributed to complex tensile and compressive waves from the opposite surfaces interacting with each other to produce numerous release waves in the mid-plane (Clauer and Lahrman, 2001).

In the experiments using multiple LSP on the thin Ti-6Al-4V section, three processing conditions were originally applied by Clauer and Lahrman

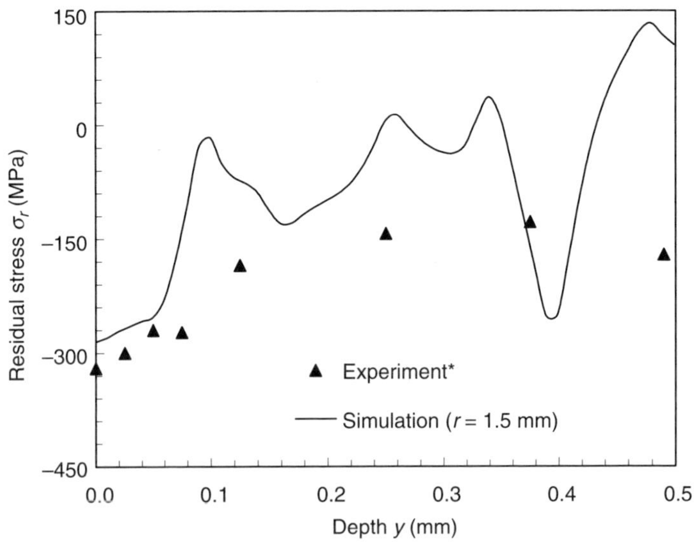

6.9 Residual stress profile in the depth of section after single LSP. (*Source*: *Clauer and Lahrman, 2001).

(2001), namely (1) low, one shot with a laser power density of 5.5 GW/cm^2; (2) intermediate, three shots with a laser power density of 5.5 GW/cm^2; (3) high, three shots with a laser power density of 10 GW/cm^2. Corresponding to these conditions, the estimated peak pressure is 4.0 GPa for the low and intermediate, but 5.5 GPa for the high. As shown in Fig. 6.10(a), there is a reasonable agreement in the trend of the residual stress profile between the experimental data and the simulated results (r_0 = 3 mm), despite the fact that the plastically affected depth associated with the compressive residual stress is difficult to determine properly.

Figure 6.10(b) shows that the predicted surface residual stress is significantly increased when using multiple LSP. The peak surface residual stress is increased by 30–40% to a level of 600 MPa from the low to intermediate and by further 40–50% to a level of 900 MPa from the intermediate to high. However, the predicted surface residual stresses at the centre and perimeter of the laser spot are clearly diversified. It is believed that, for the round laser spot typically used for LSP, the radial stress waves after release of pulse pressure may focus at the spot centre or perimeter (Masse and Barreau, 1995*a*; Braisted and Brockman, 1999). To avoid this effect, it is preferable to use a laser source that provides a square laser beam or an unsymmetrical laser beam for LSP (Masse and Barreau, 1995*a*), as the results confirmed in Chapter 5.

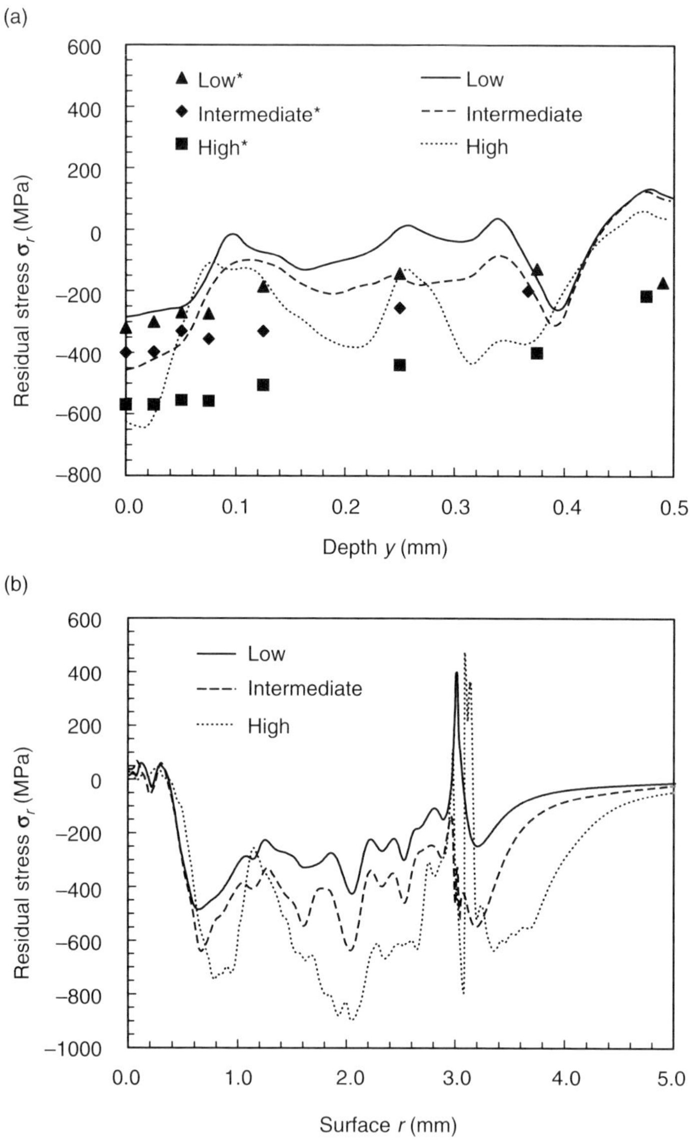

6.10 Residual stress profiles in the section after single (low) and multiple (intermediate and high) LSP with different laser power density. (a) In depth (r = 1.5 mm), (b) on surface. (*Source*: *Clauer and Lahrman, 2001).

6.5.3 Section thickness

Figure 6.11(a) shows that the predicted surface compressive residual stress at the centre or perimeter of the laser spot (r_0 = 3 mm) is slightly improved if the section thickness was increased from t = 1 mm to t = 3 mm. The

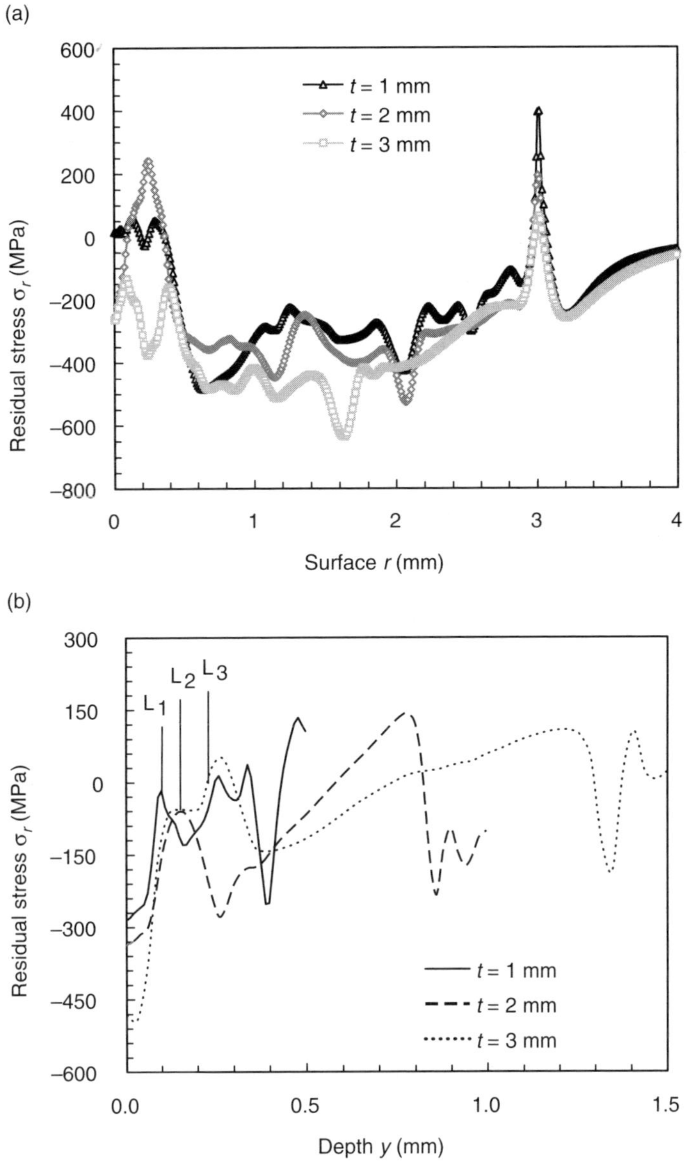

6.11 Residual stresses in sections of different thickness. (a) On the surface, (b) in depth ($r = 1.5$ mm).

maximum surface compressive residual stress for $t = 3$ mm is about 650 MPa, which is about 30% higher than for $t = 1$ mm and about 20% higher than for $t = 2$ mm. The residual stress, σ_r, distributed through the thickness of the section (at $r = 1.5$ mm), is plotted in Fig. 6.11(b) with respect to the different section thicknesses, $t = 1$, 2 and 3 mm. The magnitude of the compressive residual stress is significantly increased on the surface when the section thickness is increased from 1 to 3 mm. In addition, the plastically affected depth associated with the compressive residual stress is approximately $L_1 = 0.1$ mm for the 1 mm thick section, $L_2 = 0.15$ mm for the 2 mm thick section and $L_3 = 0.2$ mm for the 3 mm thick section. Figure 6.11(b) also shows that the residual stress changes between tension and compression in the mid-plane region of section when the thickness is increased from 1 to 3 mm. In addition, the harmful tensile residual stress in the depth can reach a peak of 130 MPa for $t = 1$ mm, 140 MPa for $t = 2$ mm and 100 MPa for $t = 3$ mm.

6.6 Summary

In this chapter, the FEM is applied to simulate the residual stress profiles of thin sections of a Ti-64Al-4V alloy impacted on both sides by single and multiple LSP. For a 1 mm thick section, the predicted residual stress profile correlates reasonably well with the available experimental data. The results from both simulations and experiments reveal that multiple LSP is a useful method for increasing the compressive residual stress in thin sections. The results from the simulations also indicate that, in the mid-plane region of thin sections, a harmful tensile residual stress field exists, but the exact magnitude of tensile residual stress and the size of region are dependent on the thickness of sections. When the section thickness is increased from 1 mm to 3 mm, the magnitude of compressive residual stress on the surface is obviously increased but the plastically affected depth is not significantly improved. Moreover, if the pulse pressure on the surface of section is increased, the surface compressive residual stress can be significantly increased. In order to achieve the best results for LSP treatment on thin sections, close attention must be given to selection of LSP conditions in relation to the section thickness, especially considering the potential harmful tensile residual stress in the middle section.

7
Simulation of laser shock peening on a curved surface

7.1 Introduction

This chapter presents two-dimensional (2D) and three-dimensional (3D) dynamic finite element (FE) simulations for the residual stress distribution induced by single and multiple LSP on the curved surfaces of a metal alloy. The predicted residual stresses for single LSP are correlated with the experimental data. The effects of LSP on the magnitude of induced residual stress in the specimen corresponding to changes in LSP conditions such as pressure, laser spot size and geometry of specimens are evaluated in detail. In particular, the effect of tensile residual stress on fatigue performance of the alloy is highlighted.

7.2 Laser shock peening model

The model is based on an experimental study on 7050-T7451 aluminium alloy bar specimens with a circular cross-section subjected to two-sided LSP (Liu *et al.*, 2002a, b). The purpose of the studies was to assess the application of LSP on critical aluminium aircraft components for improved fatigue performance. A Q-switched (Nd)-YAG laser system, used for treating these specimens, delivered 50 J per pulse with a wavelength of 1.054 μm and 25 ns in duration. Before LSP, the specimen was coated with black paint and confined by a thin layer of water to achieve a high pulse pressure without melting the specimen surface (Liu *et al.*, 2002a, b). Each specimen was simultaneously irradiated by two square laser spots of 4 × 4 mm on both sides of bar specimen. In such a confined ablation mode, the LSP process produced a peak pulse pressure of 2 GPa with a Gaussian temporal shape of a full-width at half-maximum (FWHM) of 50 ns (Peyre *et al.*, 1996).

The specimens are a cylindrical dog-bone shape, with a reduced gauge diameter of 12.0 mm (d), a length of about 90 mm (L) and two grip regions with a diameter of 24.0 mm (D) and a length of 40 mm (H), shown in Fig. 7.1(a). The schematic configuration of two-sided LSP with a confined

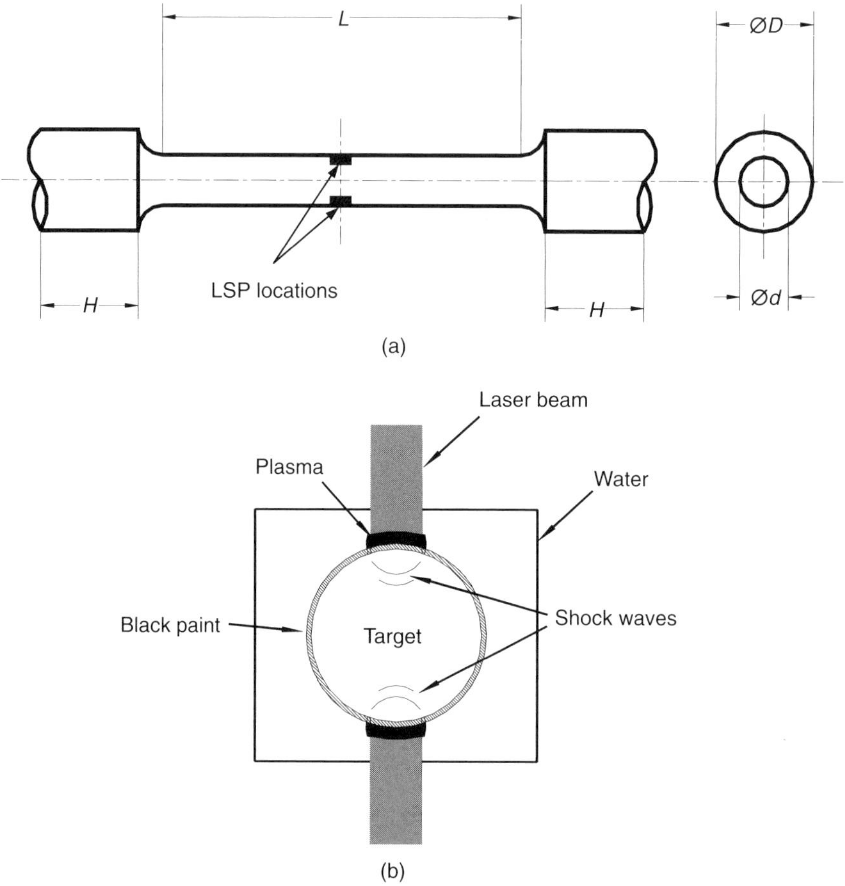

7.1 Configuration of the LSP process. (a) Specimen geometry, (b) two-sided LSP with a confined ablation mode (Ding *et al.*, 2002).

ablation mode is depicted in Fig. 7.1(b). The peened specimens were further fatigued under certain conditions (Liu *et al.*, 2002a, b).

The specimens used for the study were made from thick sections of 7050-T7451 Al alloy, which is wrought with a chemical composition of 2.07% Cu, 6.05% Zn, 2.05% Mg, 0.04% Si, 0.07% Fe, 0.04% Ti, 0.09% Zr and Al balance. The material properties of the alloy are shown in Table 7.1. It is assumed that the alloy is elastic–perfectly plastic, with isotropic and homogeneous characteristics.

7.3 Finite element models

Dynamic elastic–plastic FE models were applied in the study to predict residual stresses induced by LSP in the specimens. The simulation first

Table 7.1 Mechanical properties of 7050-T7451 aluminium alloy (Peyre *et al.*, 1998a)

Properties	Value	Unit
Density (ρ)	2830	kg/m^3
Poisson's ratio (ν)	0.33	
Elastic modulus (E)	72	GPa
Hugoniot elastic limit (HEL)	1.1	GPa
Dynamic yielding strength (σ_y^{dyn})	0.558	GPa

adopted a 2D FE model to obtain dynamic responses and residual stresses in the specimen subjected to two-sided LSP. The 2D FE model was established by addressing the cross-section of the bar specimen subjected to LSP, assuming a plane strain condition ($\varepsilon_z = 0$). As the geometry of specimen and impact loads are symmetric with respect to the x- and y-planes, a quarter of the model instead of the whole was selected for the FE analysis to ensure computational efficiency. The configuration of the 2D FE model of four-node elements with symmetric boundary conditions is depicted in Fig. 7.2(a).

A 3D FE model was also developed to simulate the process and a schematic configuration of the model with an eight-node finite element as well as infinite elements and boundary conditions is shown in Fig. 7.2(b). The infinite elements are assumed to be elastic elements, used as non-reflecting boundaries. The finite elements can undergo non-linearity with large deformation to cope with high-pressure impact. The finite element domain is adjusted in a preliminary analysis to make sure that the residual stress field is not close to the boundary between the finite and infinite elements. As a quarter of the whole model was used for the FE analyses, symmetric boundary conditions were employed on the x–0–y, y–0–z and x–0–z planes and the dimension of impact area, Z_p (= 2 mm), was equal to the half of square spot edge.

For both 2D and 3D models, it is assumed that the pulse pressure induced by the plasma is uniform over the entire surface of the laser spot. During LSP, the elastic limit of material in the direction of the shock wave propagation is defined as the Hugoniot elastic limit (HEL), as discussed in pervious chapters. It is further assumed that the material yielding follows the Von Mises criterion. The same approaches as described in Chapter 4 for simulating single and multiple LSP were applied in the evaluation. Three 3D FE models with coarse, intermediate and fine meshes were applied in order to evaluate the simulation results corresponding to the mesh refinement, defined as Ma, Mb and Mc, respectively. Their configurations are summarised in Table 7.2.

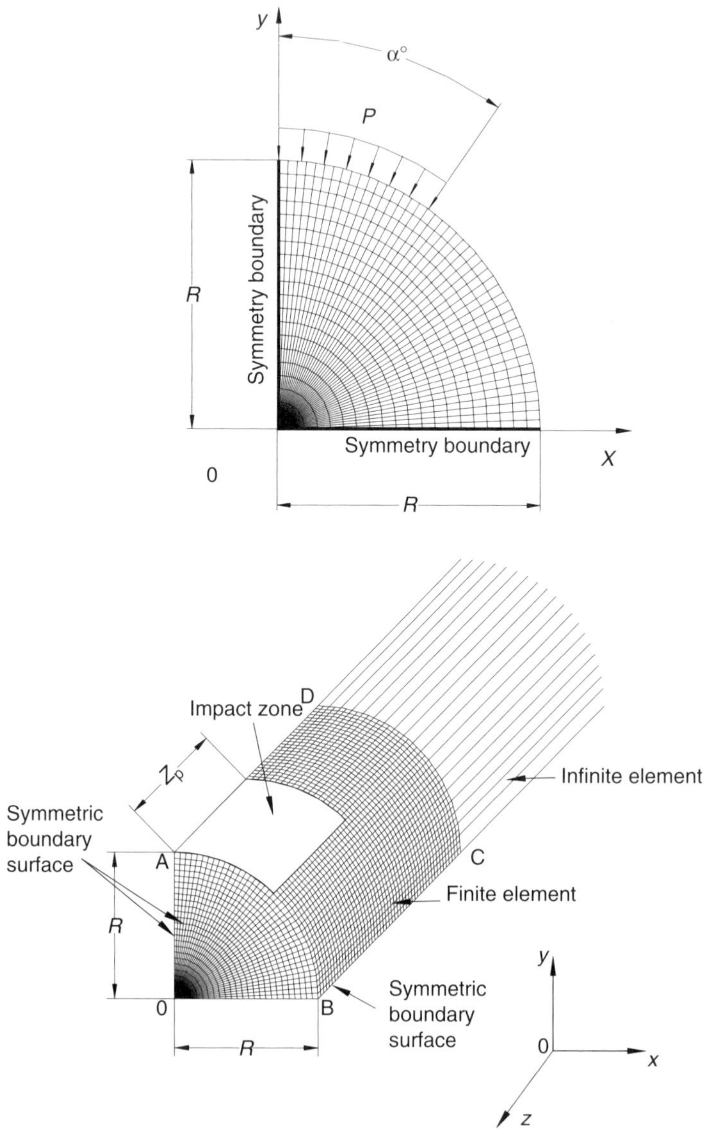

7.2 FE models with symmetric boundary conditions. (a) 2D model, (b) 3D model.

7.4 Evaluation and discussion

7.4.1 Mesh refinement

Mesh refinement is a very important factor for determining an accurate solution in FE analyses. To evaluate the FE results for mesh refinement, the

Table 7.2 Configurations of FE models of different mesh densities (Ding and Ye, 2003)

FE Model	Finite element	Infinite element	Total elements
Ma (coarse)	38220	1911	40131
Mb (intermediate)	74220	3711	77931
Mc (fine)	83220	4161	87381

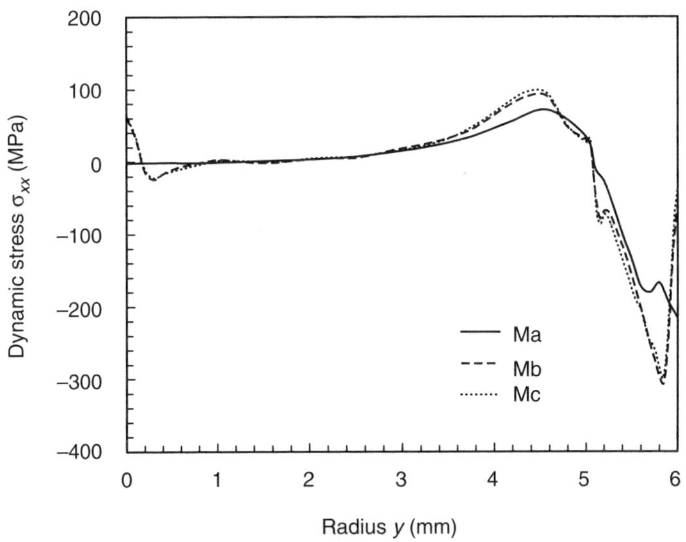

7.3 Dynamic stress (σ_{xx}) along the radius (y) for three different models, Ma, Mb and Mc, after a solution time of 100000 ns.

dynamic stresses (σ_{xx}) resulting from three 3D mesh models, Ma, Mb and Mc, along the radial direction (y-axis), are shown in Fig. 7.3. The stress profile of Ma is quite different from those of Mb and Mc, but the stress profiles between Mb and Mc are almost identical to each other. To ensure computational efficiency, the intermediate model, Mb, was selected for the further evaluation.

7.4.2 Energy dissipation

Different energy modes as the shock waves propagate through the specimen are plotted in Fig. 7.4. The plastically dissipated energy dramatically increases to 14.3 mJ in the initial period of 400 ns and remains steady at 14.6 mJ after 1160 ns, while the elastically stored energy finally drops to 2.3 mJ after 20000 ns. The steady plastically dissipated energy after 1160 ns implies that no further plastic deformation occurs in the specimen.

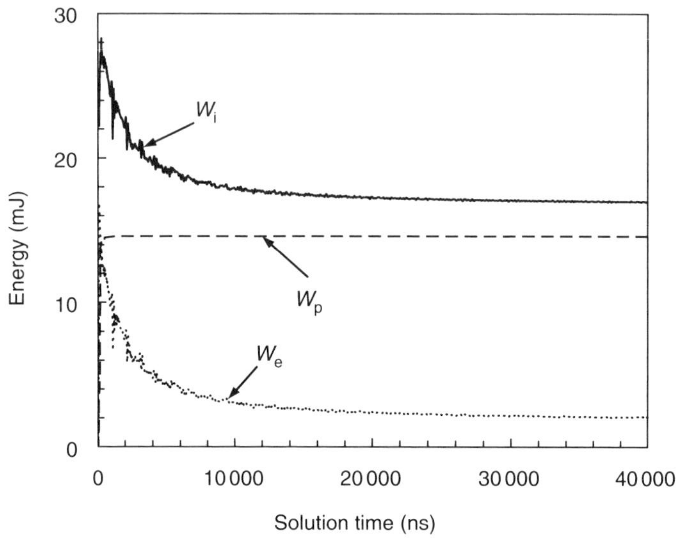

7.4 History of internal energy, elastically stored energy and plastically dissipated energy in the process.

7.4.3 Solution time

To determine a suitable solution time for the steady sate of dynamic stresses, σ_{xx} in the radial direction (y) and on the surface (AD in the axial direction), varying with the solution time, are plotted in Fig. 7.5. The dynamic stress profile changes quite clearly in the solution period from 10000 to 20000 ns, but after 40000 ns, the stress profile becomes steady, approaching the converged results. Thus, the solution time was set at 40000 ns for the FE analyses when the transient stress and strain states are imported from ABAQUS/Explicit to ABAQUS/Standard to calculate the residual stress field in static equilibrium.

7.4.4 Dynamic stress

To understand the shock wave in the specimen induced by LSP, the dynamic stresses, σ_{xx}, σ_{yy} and σ_{zz}, in the radial direction (y) of the specimen at the end of five different periods of solution time, 200, 400, 800, 1000 and 1200 ns, are depicted in Fig. 7.6. The stress waves associated with σ_{xx} and σ_{zz}, have almost the same magnitude propagating from the treated surface to the centre of cross-section. The peak magnitude of σ_{xx} (or σ_{zz}) at 200 ns is around 1.0 GPa, which is about 45% higher than that at 400 ns, about 70% higher than that at 800 ns. Compared with σ_{xx} or σ_{zz}, σ_{yy} has the highest magnitude in the process, almost always 50% higher than that of either σ_{xx} or

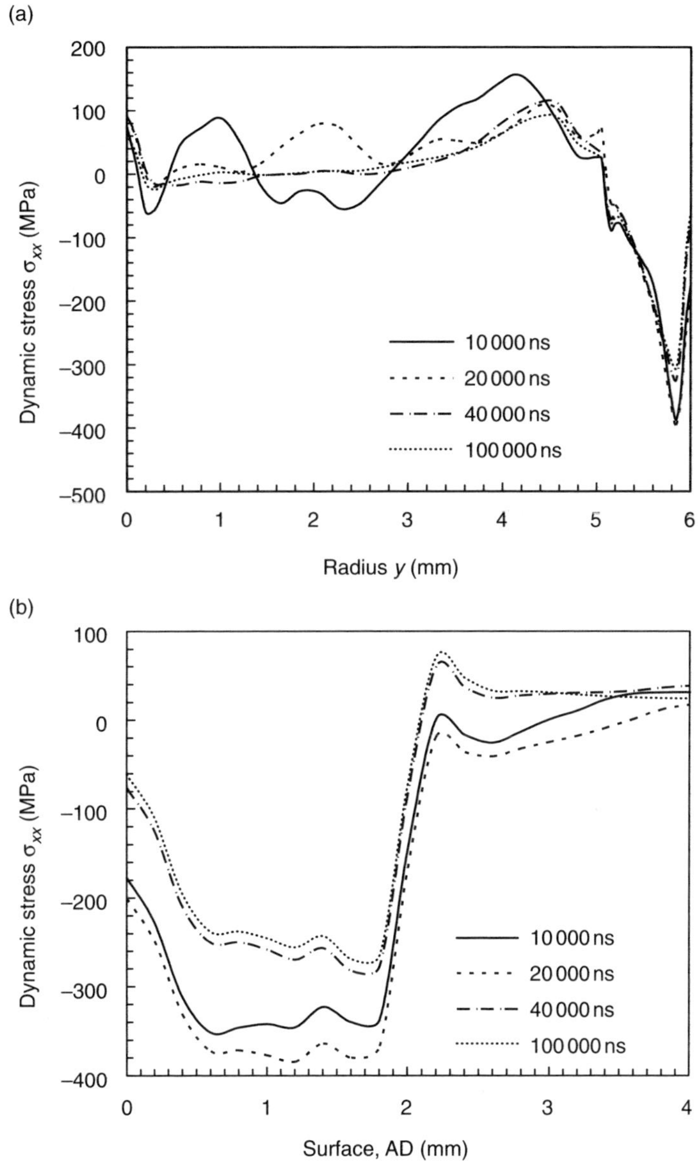

7.5 Dynamic stresses after different periods of solution time (10 000, 20 000, 40 000 and 100 000 ns). (a) σ_{xx} along the radius (y), (b) σ_{xx} on the surface.

σ_{zz}. It can be seen that, during the period from 800 to 1000 ns, the stress profiles of σ_{xx}, σ_{yy} and σ_{zz} along the propagating direction contain a leading compressive stress wave and a tailing tensile stress wave, propagating to the centre of cross-section. The stress waves from both sides of the specimen

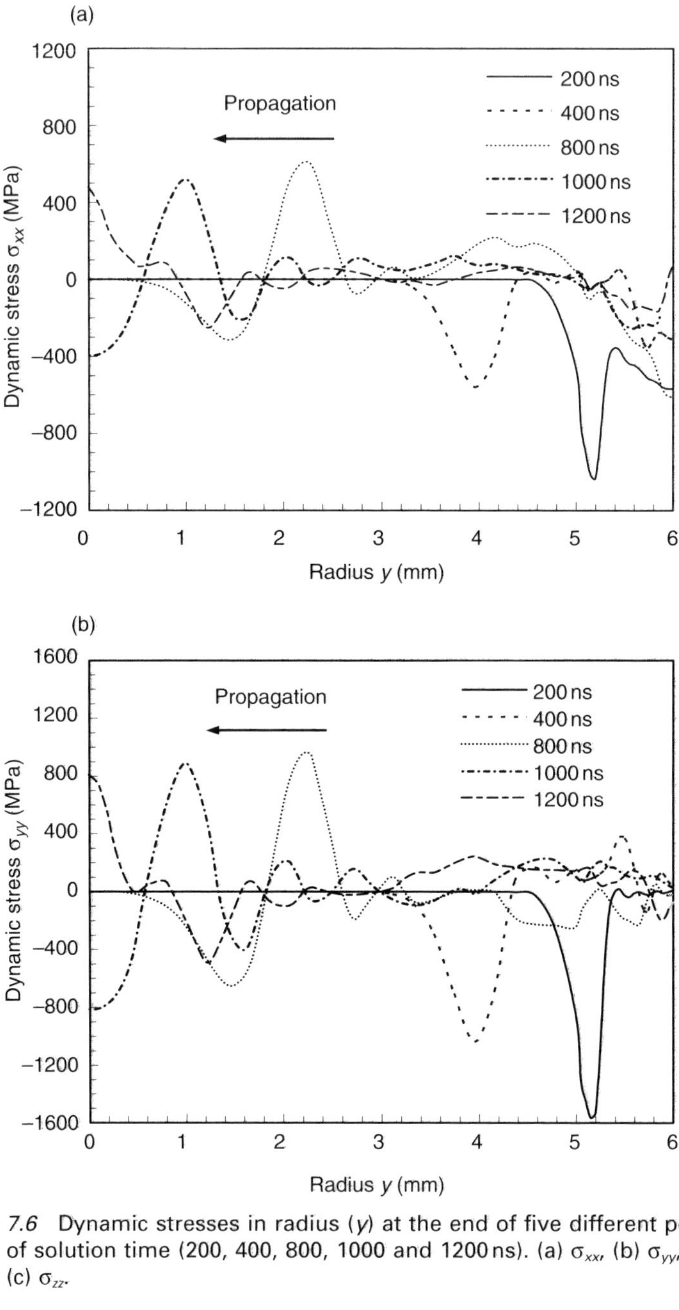

7.6 Dynamic stresses in radius (y) at the end of five different periods of solution time (200, 400, 800, 1000 and 1200 ns). (a) σ_{xx}, (b) σ_{yy}, (c) σ_{zz}.

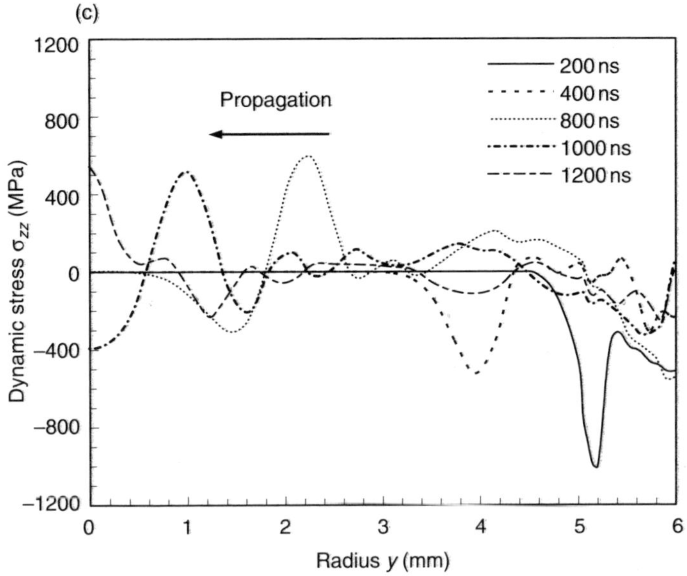

7.6 Continued

take about 1000 ns to meet at the centre where complex interaction occurs, resulting in an increase in magnitude of compressive stress but a decrease in magnitude of tensile stress. After 1000 ns, the attenuating stress waves with the leading compressive stress wave and the tailing tensile stress wave are reflected from the centre to the surface of the specimen. The fact that stresses in all directions are attenuating in magnitude with time is attributed to the plastic deformation formed in the specimen. After the shock waves have dispersed, the plastic deformation, compressive and tensile residual stresses remain. The stresses, σ_{xx} and σ_{zz}, are the dominant stresses which eventually become the compressive or tensile residual stresses, σ_x and σ_z.

7.4.5 Residual stress

Two-dimensional simulation

The residual stresses predicted by the 2D model are plotted in Fig. 7.7. Tensile residual stresses exist at the centre of cross-section and the region of tensile residual stresses forms a circle of about 1 mm in diameter around the centre of cross-section, owing to focusing of various stress waves. The compressive residual stresses are distributed along the perimeter of the cross-section. The equivalent plastic strain has penetrated to a depth of up to 1 mm from the treated surface, shown in Fig. 7.7(c). In addition, the peak plastic strain is found at the centre of cross-section.

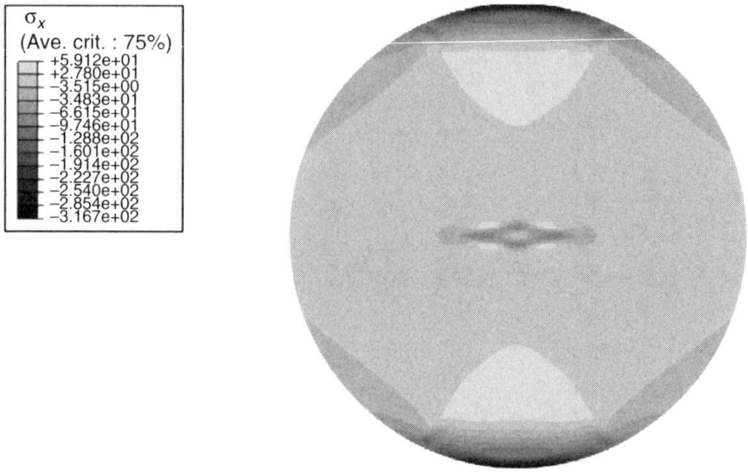

(a)

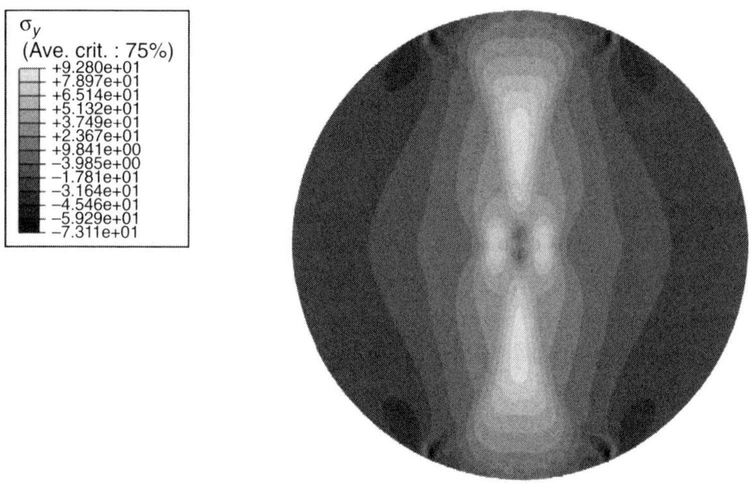

(b)

7.7 Contour of residual stress and strain in cross-section of specimen. (a) σ_x, (b) σ_y, (c) equivalent plastic strain (ε_{eq}) (Ding *et al.*, 2002).

Three-dimensional simulation

When the dynamic stresses, σ_{xx} and σ_{zz}, in the specimen reach the dynamic yielding strength of the material, they cause plastic deformation at the location. Consequently, residual stresses are generated on the surface as well as in the material. Figure 7.8 shows the residual stress in the radial direction (y) after single LSP. In this figure, the treated surface is at the right end

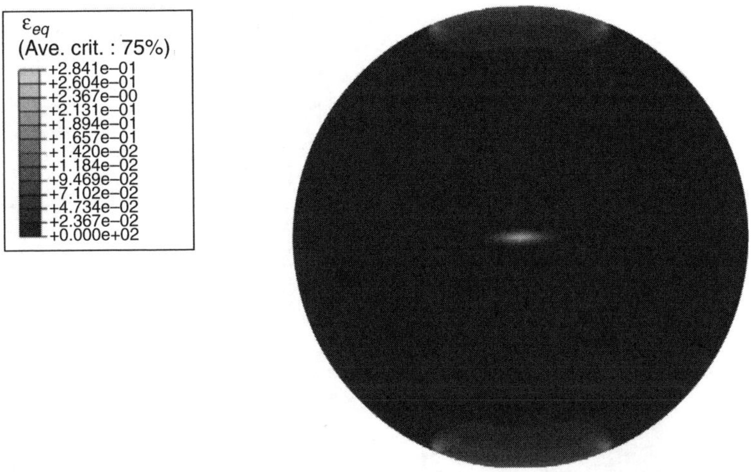

(c)

7.7 Continued

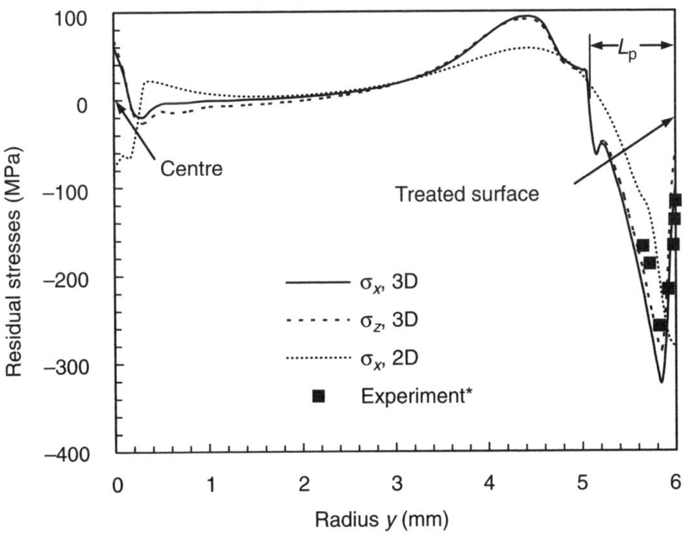

7.8 Residual stresses in radius (*y*) for single LSP. (*Source*: * Liu *et al.*, 2002a, b).

($y = 6$ mm), while the centre of bar specimen is at the left end ($y = 0$). The peak compressive residual stress, σ_x, is 320 MPa, which is about 13% higher than σ_z, but the plastically affected depth (L_p) associated with both stresses is almost the same, about 0.9 mm.

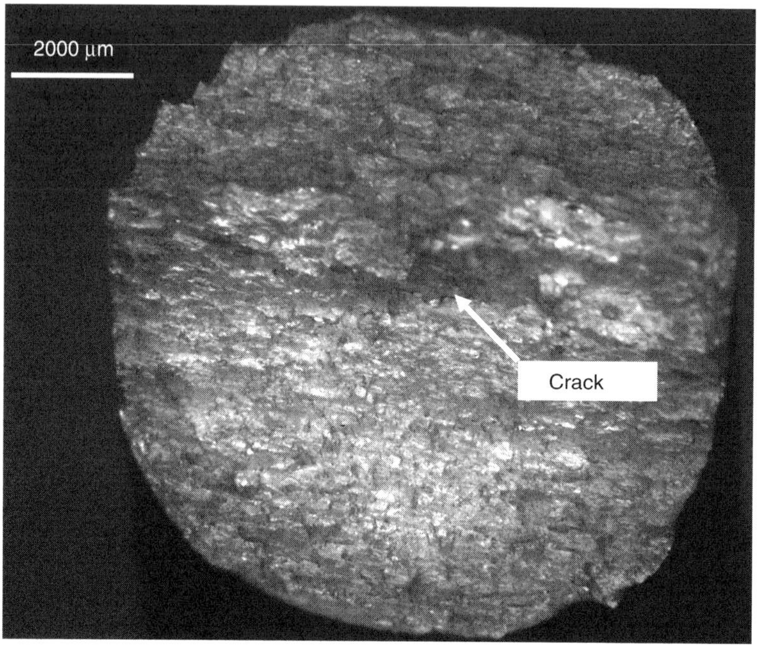

7.9 SEM micrograph of specimen transection with cracks located at the centre after a fatigue test (Liu *et al.*, 2002 a, b).

The predicted residual stresses correlate quite well with the experimental data, measured using an X-ray diffraction technique combined with an electrolytic polishing method (Liu *et al.*, 2002a, b). The simulation indicates that there are two tensile residual stress zones, at the centre and below the treated surface (about y = 4.5 mm) of the specimen, respectively. The peak tensile residual stress is about 60 MPa at the centre, while it is about 90 MPa below the treated surface (about y = 4.5 mm). In addition, the residual stress profile (σ_x) from the 2D simulation with a plane strain condition ($\varepsilon_z = 0$) has the same trend as that from the 3D FE simulation, but the magnitudes of residual stresses from the two models are quite different to each other at the centre or below the treated surface of the specimen.

A later report of fatigue tests on these peened specimens confirmed that a shorter-than-usual fatigue life was obtained, with internal fatigue cracking or fracturing near the centre of specimen (Liu *et al.*, 2002a, b), which can be attributed to the tensile residual stresses developed by LSP. The typical surface morphology of cross-sectional failure near the centre of specimen after the fatigue tests is shown in Fig. 7.9 (Liu *et al.*, 2002a, b). There is a fatigue crack across the centre part of specimen and the crack also involves multiple cracking, shown in detail in Fig. 7.10.

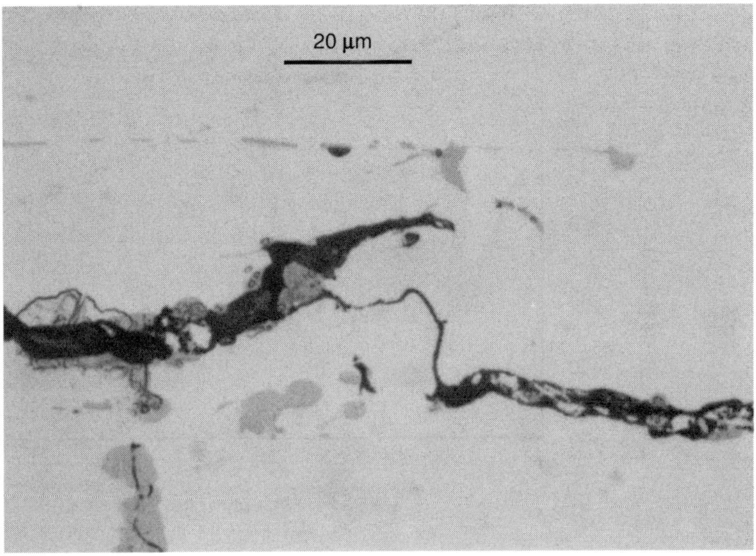

7.10 SEM micrograph of cracks at the centre of specimen (Liu *et al.*, 2002a, b).

7.4.6 Parameter study

Multiple laser shock peening

The predicted surface residual stress distributions in the axial direction resulting from multiple LSP at the same spot are shown in Fig. 7.11. The peak surface compressive residual stress (σ_x) for single LSP is about 300 MPa and the corresponding plastically affected depth is 0.9 mm. Compared with the single impact, the peak surface residual stress and plastically affected depth after four impacts on the same spot are significantly increased, by 33% to 400 MPa and by 39% to 1.25 mm, respectively. It was found that the plastically affected depth increased gradually with the number of impacts on the same spot, while the surface residual stress gradually achieved a saturated level (Clauer, 1996; Peyre *et al.*, 1996).

Geometry of specimen

The simulations were performed to evaluate the residual stress field in the specimens with changes in the diameter of the specimen, using the same LSP conditions. Four different diameters of specimen were assumed for the 3D model, 10, 12, 16 and 20 mm, respectively. Figure 7.12(a) indicates that the surface residual stress profiles are slightly changed when the diameter of specimen is increased from 10 to 16 mm, while there is a clear increase

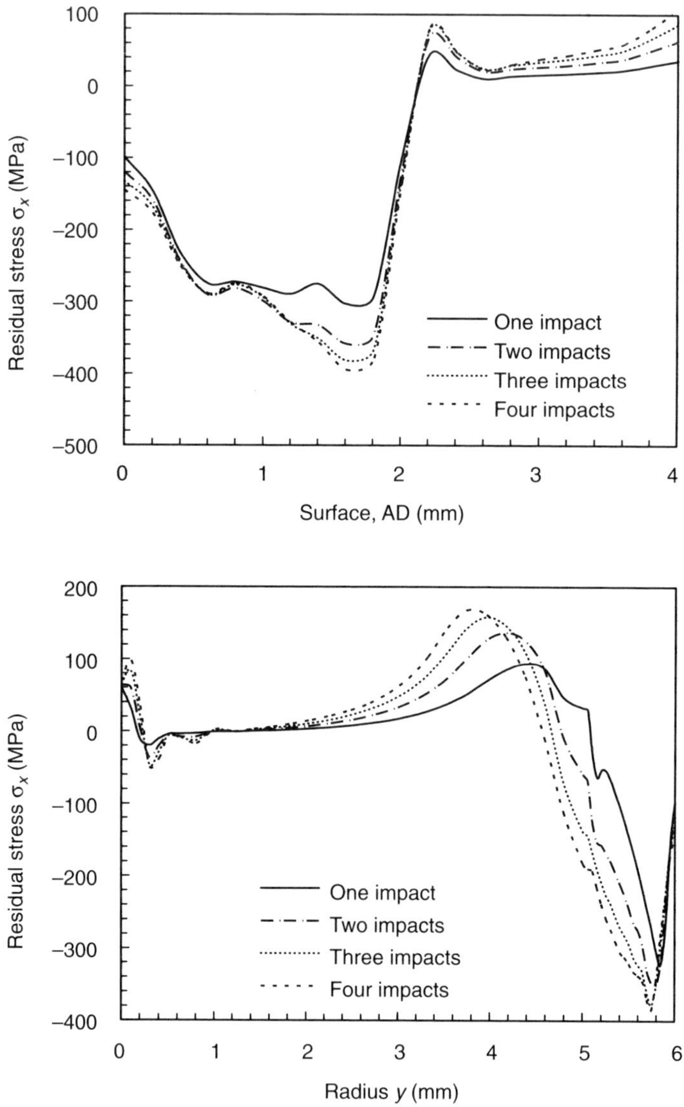

7.11 Residual stress, σ_x, after multiple LSP.

in the peak surface compressive residual stress for the specimen of a diameter of 20 mm. The plastically affected depth in all four specimens remains almost unchanged at 0.9 mm, but there is no tensile residual stress at the centre of specimen when the diameter is 16 mm or 20 mm, shown in Fig. 7.12(b).

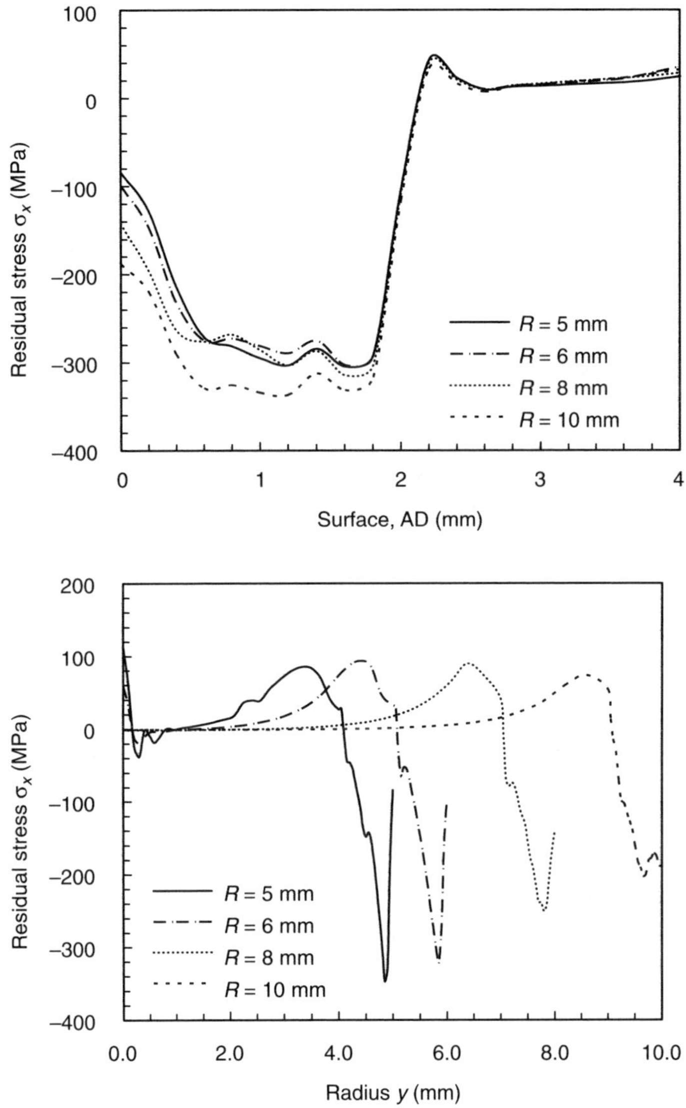

7.12 Residual stress, σ_x, for specimens of different diameters.

Laser spot size

The residual stress field in the specimen with respect to changes in laser spot size was evaluated by 3D FE analyses. The half of square laser spot edge (Z_p), was set to 1, 2 and 3 mm, respectively. The results in Fig. 7.13 show that the residual stress profiles on the surface (AD) in the axial direction

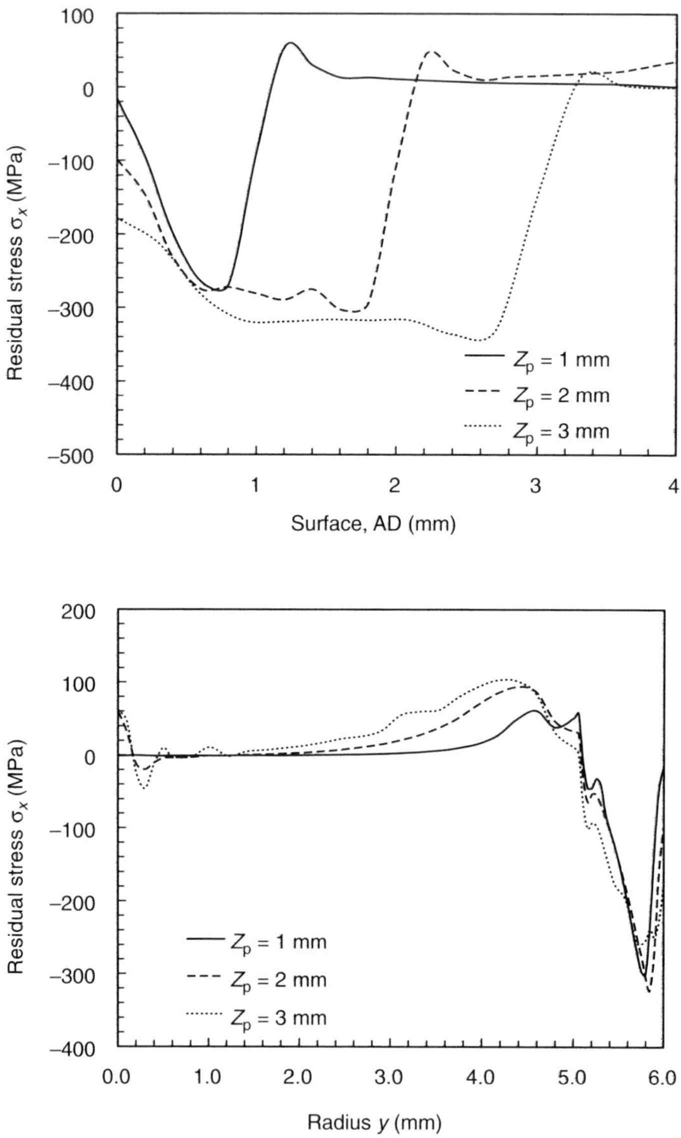

7.13 Residual stress, σ_x, for different laser spot sizes after single LSP.

change significantly when Z_p is increased from 1 to 3 mm. However, the plastically affected depth is almost unchanged at 0.9 mm as a result of increasing the laser spot size, but there is no tensile residual stress at the centre for $Z_p = 1$ mm, shown in Fig. 7.13(b).

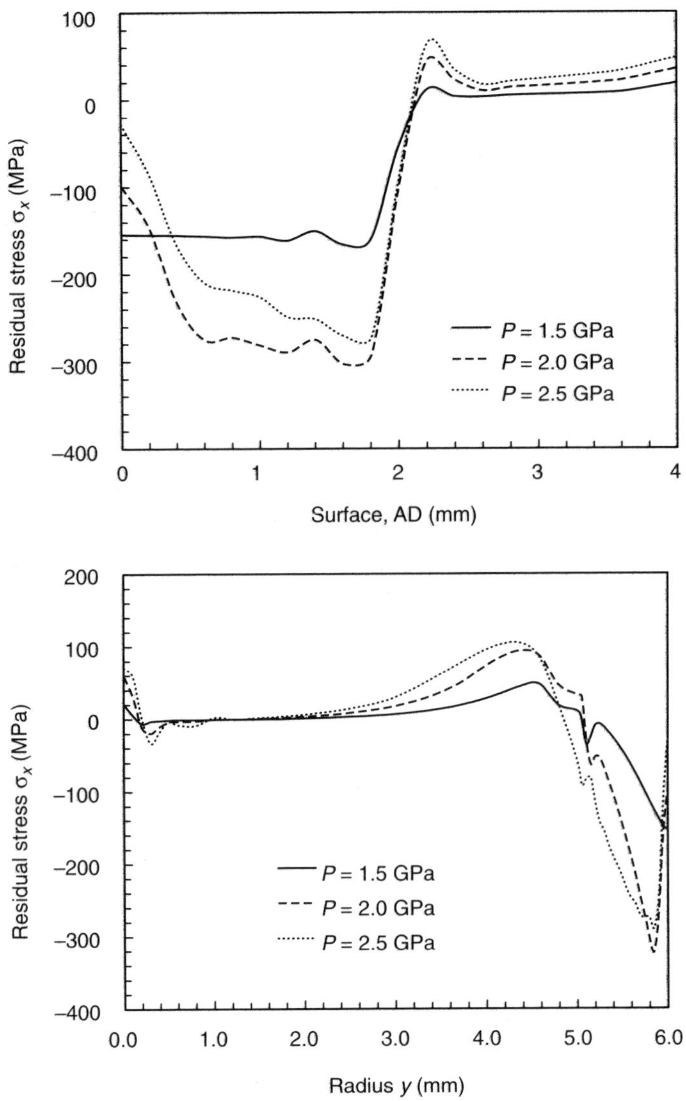

7.14 Residual stress, σ_x, for different peak pressures after single LSP.

Impact pressure

The residual stress field with respect to changes in impact pressure was also evaluated by 3D FE analyses. As evaluated in previous chapters, the pressure induced by LSP is a function of the laser power density (Fabbro *et al.*, 1998; Montross *et al.*, 2002). An increase in laser power density can result in an increase in the magnitude of pulse pressure on the specimen surface

(Fabbro *et al.*, 1998; Montross *et al.*, 2002; Devaux *et al.*, 1993). Consequently, the peak pressure used for the FE analyses was set to be 1.5, 2.0 and 2.5 GPa, as a result of increasing the laser power density. In addition, the simulation for each case was accomplished using the same pressure pulse duration of 100 ns (FWHM = 50 ns) for the pressure–time history and the specimen is 12 mm in diameter. The results in Fig. 7.14(a) show that the residual stress profiles are clearly changed when using the different peak pressures. The peak surface compressive residual stress is about 160, 300 and 280 MPa when the peak pressure is 1.5, 2.0, and 2.5 GPa, respectively. The plastically affected depth is slightly increased when the peak pressure is increased from 1.5 to 2.5 GPa, shown in Fig. 7.14(b). Increasing the peak pressure can result in an increase in peak surface residual stress, but when the peak pressure is more than twice the HEL, that is, 2.2 GPa in the present case, a saturated level of surface residual stress is found. The saturation in the surface residual stress can be attributed to the magnitude of dynamic yield strength of the material (Fabbro *et al.*, 1998; Peyre *et al.*, 1996). Although increasing the peak pressure leads to an increase in the magnitude of surface compressive residual stress, tensile residual stresses of a high magnitude are observed at the centre of the specimen.

7.5 Summary

The residual stress in fatigue testing specimens of a 7075-T7451 aluminium alloy with a circular cross-section induced by single and multiple LSP was evaluated using both 2D and 3D FE models. For single LSP, the simulated residual stresses correlate quite well with the experimental data. As a result of single LSP, the peak surface compressive residual stress reaches a level of 300 MPa with a plastically affected depth of 0.9 mm. The simulated results reveal that the compressive residual stress is increased in magnitude and driven deeper below the surface with multiple LSP on the same spot. An increase in surface residual stress can also be achieved when the diameter of the specimen and the laser spot size are increased. However, the plastically affected depth is almost independent of the diameter of the specimen and the laser spot size. When a high pulse pressure is used for the LSP process, the magnitude of both surface compressive residual stress and plastically affected depth is increased, but there is a saturated level for the surface residual stress when the peak pressure is more than twice the HEL. The simulation also reveals that increasing the diameter of the specimen or reducing the laser spot size or the peak pressure are effective ways of minimising harmful tensile residual stress at the centre of specimens.

References

ABACUS (1998). *ABAQUS/Explicit and ABAQUS/Standard Version 6.2, 1998*. Hibbitt, Karlsson & Sorensen, USA.

ACHENBACH, J.D. (1973). *Wave Propagation in Elastic Solids.* American Elsevier Publishing Company, New York.

AL-OBAID Y.F. (1990). Three-dimensional dynamic finite element analysis for shot-peening mechanics. *Computer and Structures*, **36**, 681–9.

AL-OBAID Y.F. (1991). The automated simulation of dynamic non-linearity to shot-peening mechanics. *Computer and Structures*, **40**, 1451–60.

ALTSHULIN S., ZAHAVI A.R. and NADIV S. (1990). The interaction between a pulsed laser beam and a steel surface. *Materials Science*, **25**, 2259–63.

ANDERHOLM N.C. (1970). Laser-generated stress waves. *Applied Physics Letters*, **16**, 113–14.

AUROUX E., BOUSTIE M., ROMAIN J.P., BERTHEAU D., PEYRE P., BERTHE L. and BARTNICKI E., 2001. Debonding study of Ni-base substrate/Pt coatings interfaces using laser shock waves: characterization of the targets and experimental study. *Surface and Coatings Technology*, **138**, 269–77.

BALLARD P. (1991). *Residual Stresses Induced by Rapid Impact – Applications of Laser Shocking*. Doctorial Thesis, Ecole Polytechnique, France.

BALLARD P., FOURNIER J., FABBRO R. and FRELAT J. (1991). Residual stresses induced by laser-shocks. *Journal de Physique IV*, **C3**, 487–94.

BALLARD P., FRELAT J., ROUGIER Y. and GIRARDOT D. (1995). Plastic strain and residual stress fields induced by a homogeneous impact over the boundary of an elastic–plastic half-space. *European Journal of Mechanics A/Solids*, **14**, 1005–16.

BANAS G., ELSAYED-ALI H.E., LAWRENCE JR.F.V. and RIGSBEE J.M. (1990a). Laser shock-induced mechanical and microstructural modification of welded maraging steel. *Applied Physics*, **67**, 2380–4.

BANAS G., LAWRENCE JR.F.V., RIGSBEE J.M. and ELSAYED-ALI H.E. (1990b). Laser shock hardening of welded maraging steel. *Surface Engineering*, **67**, 280–90.

BATHE K.J., (1996). *Finite Element Procedure.* Prentice-Hall, Inc. Upper Saddle River, New Jersey.

BERTHE L., FABBRO R., PEYRE P., TOLLIER L. and BARTNICKI E. (1997). Shock waves from a water-confined laser-generated plasma. *Journal of Applied Physics*, **82**, 2826–32.

BERTHE L., FABBRO R., PEYRE P. and BARTNICKI E. (1999). Wavelength dependent laser shock-wave generation in the water-confinement regime. *Applied Physics*, **85**, 7552–5.

BOLGER J.A., MONTROSS C.S. and RODE A.V. (1999). Shock waves in basalt rock generated with high-powered lasers in a confined geometry. *Journal of Applied Physics*, **86**, 5461–6.

BRAISTED W. and BROCKMAN R. (1999). Finite element simulation of laser shock peening. *International Journal of Fatigue*, **21**, 719–24.

BROWN A.S. (1998). A shocking way to strengthen metal. *Aerospace America*, **36**(4), 21–3.

CAI L. and ZHANG Y. (1996). Study of laser parameters optimum of laser shocking against fatigue and fracture of metal. *Chinese Journal of Lasers*, **23A**, 1117–20.

CASARCIA D.A., COWIE W.D. and MANNAVA S. (1996). *Laser Shock Peened Bearings.* US Patent 5584586, General Electric Company.

CHU J.P., RIGSBEE J.M., BANAS G., LAWRENCE JR.F.V. and ELSAYED-ALI H.E. (1995). Effects of laser-shock processing on the microstructure and surface mechanical properties of hadfield manganese steel. *Metallurgical and Materials Transactions A*, **26A**, 1507–17.

CHU J.P., RIGSBEE J.M., BANAS G. and ELSAYED-ALI H.E. (1999). Laser-shock processing effects on surface microstructure and mechanical properties of low carbon steel. *Materials Science and Engineering*, **A260**, 260–8.

CLAUER A.H. (1996). Laser shock peening for fatigue resistance. *Proceedings of Surface Performance of Titanium*, Gregory J.K., Rack H.J. and Eylon D. (eds), TMS, Warrendale, PA. The Metal Society of AIME, 217–30.

CLAUER A.H. and FAIRAND B.P. (1979). Interaction of laser-induced stress waves with metals. *ASM Conference Applications of Laser in Material Processing*, Washington DC, ASM International, Material Park, OH 44073-0002, 1–22.

CLAUER A.H. and KOUCKY J.P. (1991). Laser shock processing increases the fatigue life of metal parts. *Materials and Processing*, **6**, 3–5.

CLAUER A.H. and LAHRMAN D.F. (2001). Laser shock processing as a surface enhancement process. *Engineering Materials*, **197**, 121–44.

CLAUER A.H., FAIRand B.P. and WILCOX B.A. (1976). Laser shock hardening of weld zones in aluminium alloys. *Metallurgical Transactions A*, **8A**, 1871–6.

CLAUER A.H., FAIRand B.P. and WILCOX B.A. (1977). Pulsed laser induced deformation in an Fe-3 wt Pct Si alloy. *Metallurgical Transactions A*, **8A**, 119–25.

CLAUER A.H., HOLBROOK J.H. and FAIRAND B.P. (1981). Effects of laser induced shock waves on metals, *Shock Waves and High-Strain-Rate Phenomena in Metals*, Meyers M.A. and Murr L.E. (eds), Plenum Publishing Corporation, New York, 675–703.

CLAUER A.H., WALTERS C.T. and FORD S.C. (1983). The effects of laser shock processing on the fatigue properties of 2024-T3 aluminum. *Lasers in Materials Processing*, Metzbower E.A. (ed), American Society for Metals, Metals Park, Ohio, 7–22.

CLAUER A.H., DULANEY J.L., RICE R.C. and KOUCKY J.R. (1992). Laser shock processing for treating fastener holes in aging aircraft. *Proceedings of the International Workshop on Structural Integrity of Aging Airplanes*, Atluri S.N., Harris C.E., Hoggard A., Miller N. and Sampath S.G. (eds), Atlanta Technology, Atlanta, 350–61.

CLAUER A.H., DULANEY J.L., TOLLER S.M. and WALTERS C.T. (1998a). *Oblique Angle Laser Shock Processing.* Canada Patent 2226444, LSP Technologies.

CLAUER A.H., DULANEY J.L., TOLLER S.M. and WALTERS C.T. (1998b). *Hidden Surface Laser Shock Processing.* Canada Patent 2226451, LSP Technologies.

COOK R.D., MALKUS D.S. and PLESHA M.E. (1989). *Concepts and Applications of Finite Element Analysis*, John Wiley & Sons, Canada.

COTTET F. and BOUSTIE M. (1989). Spallation studies in aluminum targets using shock waves induced by laser irradiation at various pulse durations. *Journal of Applied Physics*, **66**, 4067–73.

COTTET F., MARTY L., HALLOUIN M., ROMAIN J.P., VIRMONT J., FABBRO R. and FARAL B. (1988). Two-dimensional study of shock breakout at the rear face of laser irradiated metallic targets. *Journal of Applied Physics*, **64**, 4474–7.

COUTURIER S., M. DE RESSEDUIER, HALLOUIN M., ROMAIN J.P. and BAUER F. (1996). Shock profile induced by short laser pulses. *Journal of Applied Physics*, **79**, 9338–42.

DAI S.J., LIU F.R., YANG X. and ZOU H.C. (1997). Effects of laser shock processing on material's properties. *Transactions of the Nonferrous Metals Society of China*, **7**, 120–4.

DANE C.B., HACKEL L.A., DALY J. and HARRISON J. (1997). High laser power for peening of metals enabling production technology. *Advanced Aerospace Materials and Processes Conference '98*, Tysons Corner, Virginia, June 15–18, 1998.

DE RESSÉGUIER T., COUTURIER S., DAVID J. and NIÉRAT G. (1997). Spallation of metal targets subjected to intense laser shocks. *Journal of Applied Physics*, **82**(5), 2617–32.

DESAI, C.S. and SIRIWARDANE, H.J. (1984). *Constitutive Laws For Engineering Materials with Emphasis on Geologic Materials*, Prentice-Hall, Englewood Cliffs, New Jersey.

DEVAUX D., FABBRO R. and VIRMONT J. (1991). Generation of shock waves by laser–matter interaction in confined geometries. *Journal De Physique IV*, **1**(C7), 179–82.

DEVAUX D., FABBRO R., TOLLIER L. and BARTNICKI E. (1993). Generation of shock waves by laser-induced plasma in confined geometry. *Journal of Applied Physics*, **74**, 2268–73.

DING K. (2003). *A Study on Finite Element Simulation of Laser Shock Peening Processes in Metallic Materials*, PhD Dissertation, University of Sydney.

DING K. and YE L. (2003a). FEA simulation of two-sided laser shock peening of thin sections of a Ti-6Al-4V alloy, *Surface Engineering*, **19**(2), 127–33.

DING K. and YE L. (2003b). 3-D dynamic finite element analysis of multiple laser shock peening processes, *Surface Engineering*, **19**(5), 351–8.

DING K., YE L. and LIU Q.C. (2002). Finite element simulation of laser shock peening on an aluminium alloy, in *Proceedings of the Third Australian Congress on Applied Mechanics*, Zhang L., Tong L. and Gal J. (eds), World Scientific, Sydney, Australia, 605–11.

FABBRO R., FOURNIER J., BALLARD P., DEVAUX D. and VIRMONT J. (1990). Physical study of laser-produced plasma in confined geometry. *Journal of Applied Physics*, **68**, 775–84.

FABBRO R., PEYRE P., BERTHE L. and SCHERPEREEL X. (1998). Physics and applications of laser-shock processing. *Laser Application*, **10**, 265–79.

FAIRand B.P. and CLAUER A.H. (1977). Use of laser generated shocks to improve the properties of metals and alloys. *Industrial Applications of High Power Laser Technology*, **86**, 112–19.

FAIRAND B.P. and CLAUER A.H. (1978). Laser generation of high-amplitude stress waves in materials. *Journal of Applied Physics*, **50**, 1497–502.

FAIRAND B.P., WILCOX B.A., GALLAGHER W.J. and WILLIAMS D.N. (1972). Laser shock-induced microstructural and mechanical property changes in 7075 aluminum. *Journal of Applied Physics*, **43**, 3893–5.

FAIRAND B.P., CLAUER A.H., JUNG R.G. and WILCOX B.A. (1974). Quantitative assessment of laser-induced stress waves generated at confined surfaces. *Journal of Applied Physics*, **25**, 431–3.

FERRIGNO S.J., COWIE W.D. and MANNAVA S. (1998). *Laser Shock Peening for Gas Turbine Engine Weld Repair.* US Patent 5 846 057, General Electric Company.

FERRIGNO S.J., MCALLISTER K.G. and MANNAVA S. (2001). *Laser Shock Peened Gas Turbine Engine Seal Teeth.* US Patent 6 200 689, General Electric Company.

FOUQUET F., RENAUD L., MILLET J.P. and MAZILLE H. (1991). Laser surface alloying of a mild steel for corrosion resistance improvement. *Journal de Physique IV*, **1**(C7), 69–72.

GERLAND M. and HALLOUIN M. (1994). Effect of pressure on the microstructure of an austenitic stainless steel shock-loaded by very short laser pulses. *Materials Science*, **29**, 345–51.

GERLAND M., HALLOUIN M. and PRESLES H.N. (1992). Comparison of two new surface treatment processes, laser-induced shock waves and primary explosive: application to fatigue behaviour. *Materials Science and Engineering*, **A156**, 175–82.

GREVEY D., MAIFFREDY L. and VANNES A.B. (1992). Laser shock on a TRIP alloy: mechanical and metallurgical consequences. *Materials Science*, **27**, 2110–16.

GRIFFIN R.D., JUSTUS B.L., CAMPILLO A.J. and GOLDBERG, L.S. (1986). Interferometric studies of the pressure of a confined laser-heated plasma. *Journal of Applied Physics* **59**, 1968–71.

GUO D.H., WU H.X., WANG S.B., HONG X., WANG J., DAI Y.S., XIA X.P., ZHANG Y.K., ZHANG H., TANG Y.X. and YU C.Y. (1999). Study of mechanism of strengthening materials with laser shock processing. *Science in China*, **42**, 287–93.

HOFFMAN C.G. (1974). Laser–target interactions. *Journal of Applied Physics*, **45**, 2125–8.

HOFFMAN, O. and SCAHS, G. (1953). *Introduction to the Theory of Plasticity for Engineers*, McGraw-Hill Book Company, New York.

HONG X., WANG S.B., GUO D.H., WU H.X., WANG J., DAI Y.S., XIA X.P. and XIE Y.N. (1998). Confining medium and absorptive overlay: their effects on a laser-induced shock wave. *Journal of Optics and Lasers in Engineering*, **29**, 447–55.

JOHNSON J.N. and ROHDE R.W. (1971). Dynamic deformation twining in shock-loaded iron. *Journal of Applied Physics*, **42**, 4171–3.

JOHNSON, W. (1972). *Impact Strength of Materials*, Edward Arnold, London.

KACHANOV, L.M. (1971). *Foundations of Theory of Plasticity*, North-Holland, Amsterdam.

KHABOU M.T., CASTEX L. and INGLEBERT G. (1990). The effect of material behaviour law on the theoretical shot peening results. *European Journal Mechanics A/Solids*, **9**, 537–49.

KOLSKY, H. (1953). *Stress Waves in Solids*, Oxford University Press, London.

LI J.K., MEI Y. and WANG D. (1991). Mechanical approach to the residual stress field induced by shot peening. *Materials Science and Engineering*, **A147**, 167–73.

LING P. and WIGHT C.A. (1995). Laser-generated shock waves in thin films of energetic materials. *Journal of Applied Physics*, **78**, 7022–5.

LIU Q.C., BARTER S. and CLARK G. (2002a). Internal cracking during surface treatment of 7050-T74 aluminium alloy using laser shock peening. *Structural Integrity and Fracture*, Dyskin A.V., Hu X.Z. and Sahouryeh E. (eds), Swets & Zeitlinger, Perth, Australia, 177–82.

LIU Q.C., BARTER S.A. and CLARK G., (2002b). Comparison of shot peening and laser shock peening techniques in enhancing fatigue life of 7xxx series aluminium alloys, in *Proceedings of the Third Australian Conference of Applied Mechanics 2002 (ACAM 2002)*, Zhang L., Tong L. and Gal J. (eds), Sydney, Australia, World Scientific, 153–8.

LU J. (ed) (1996). *Handbook of Measurement of Residual Stresses*, Society for Experimental Mechanics, Bethel, Connecticut.

MALLOZI P.I. and FAIRand B.P. (1974). *Altering Material Properties.* US Patent 3 850 698. USA, Industrial Materials Limited.

MANNAVA S. (1998). *On the Fly Laser Shock Peening.* US Patent 5 756 965, General Electric Company.

MANNAVA S. and COWIE W.D. (1996). *Technique to Prevent or Divert Cracks.* US Patent 5 569 018, General Electric Company.

MANNAVA S. and FERRIGNO S.J. (1997). *Laser Shock Peening for Gas Turbine Engine Vane Repair.* US Patent 5 675 892, General Electric Company.

MANNAVA S., MCDANIEL A.E., COWIE W.D., HALILA H., RHODA J.E. and GUTKNECHT J.E. (1997). *Laser Shock Peened Gas Turbine Engine Fan Blade Edges.* US Patent 5 591 009, General Electric Company.

MASSE J.E. and BARREAU G. (1995a). Laser generation of stress waves in metal. *Surface and Coatings Technology*, **70**, 231–4.

MASSE J.E. and BARREAU G. (1995b). Surface modification by laser induced shock waves. *Surface Engineering*, **11**, 131–2.

MEGUID S.A., SHAGAL G. and STRANART J.C. (1999a). Finite element modelling of shot-peening residual stresses. *Materials Processing Technology*, **92–93**, 401–4.

MEGUID S.A., SHAGAL G., STRANART J.C. and DALY J. (1999b). Three-dimensional dynamic finite element analysis of shot-peening induced residual stresses. *Finite Elements in Analysis and Design*, **31**, 179–91.

MONTROSS C.S. and FLOREA V. (2001). The influence of coatings on subsurface mechanical properties of laser peened 2011-T3 aluminium. *Journal of Materials Science*, **36**, 1801–7.

MONTROSS C.S., FLOREA V. and BOLGER J.A. (1999). Laser-induced shock wave generation and shock wave enhancement in basalt. *Rock Mechanics and Mining Sciences*, **36**, 849–55.

MONTROSS C.S. and FLOREA V., SWAIN M.V. (2000). Subsurface properties of laser peened 6061-T6 Al weldments. *Surface Engineering*, **16**, 116–21.

MONTROSS C.S., WEI T., YE L., CLARK G. and MAI Y.M. (2002). Laser shock processing and its effects on microstructure and properties of metal alloys: a review. *International Journal of Fatigue*, **24**, 1021–36.

NOACK J., HAMMER D.X., NOOJIN G.D., ROCKWELL B.A. and VOGEL A. (1998). Influence of pulse duration on mechanical effects after laser-induced breakdown in water. *Journal of Applied Physics*, **83**, 7488–95.

NOWACKI, W.K. (1978). *Stress Waves in Non-elastic Solids*, Pergamon Press, UK.

O'KEEFE J.D. and SKEEN C.H. (1972). Laser-induced stress-wave and impulse augmentation. *Journal of Applied Physics*, **21**, 464–6.

OBATA M., SANO Y., MUKAI N., YODA M., SHIMA S. and KANNO M. (1999). Effect of laser peening on residual stress and stress corrosion cracking for type 304 stainless steel. *The 7th International Conference on Shot Peening (ICSP-7)*, Institute of Precision Mechanics, Warsaw.

ORAN, E.S. and BORIS, J.P. (2001). *Numerical Simulation of Reactive Flow*, Second Edition, Cambridge University Press, UK.

PEYRE P. and FABBRO, R. (1995a). Electromagnetic gauge study of laser-induced shock waves in aluminium alloys. *Journal De Physique III France*, **5**, 1953–64.

PEYRE P. and FABBRO R. (1995b). Laser shock processing: a review of the physics and applications. *Optical and Quantum Electronics*, **27**, 1213–29.

PEYRE P., MERRIEN P., LIEURADE H.P. and FABBRO R. (1995). Laser induced shock waves as surface treatment for 7075-T7351 aluminium alloy. *Surface Engineering*, **11**, 47–52.

PEYRE P., FABBRO R., MERRIEN P. and LIEURADE H.P. (1996). Laser shock processing of aluminium alloys: application to high cycle fatigue behaviour. *Materials Science and Engineering*, **A210**, 102–13.

PEYRE P., BERTHE L., SCHERPEREEL X., FABBRO R. and BARTNICKI E. (1998a). Experimental study of laser-driven shock waves in stainless steels. *Applied Physics*, **84**, 5985–92.

PEYRE P., BERTHE L., SCHERPEREEL X. and FABBRO R. (1998b). Laser-shock processing of aluminium-coated 55C1 steel in water-confinement regime, characterization and application to high-cycle fatigue behaviour. *Materials Science*, **33**, 1421–9.

PEYRE P, SCHERPEREEL X., BERTHE L. and FABBRO R. (1998c). Current trends in laser shock processing. *Surface Engineering*, **14**, 377–80.

PEYRE P., BERTHE L., FABBRO R. and SOLLIER A. (2000a). Experimental determination by PVDF and EMV techniques of shock amplitudes induced by 0.6–3 ns laser pulses in a confined regime with water. *Journal of Physics D: Applied Physics*, **33**, 498–503.

PEYRE P., SCHERPEREEL X., BERTHE L., CARBONI C., FABBRO R., BERANGER G. and LEMAITRE C. (2000b). Surface modifications induced in 316L steel by laser peening and shot-peening – influence on pitting corrosion resistance. *Materials Science and Engineering*, **A280**, 294–302.

PEYRE P., BRAHAM C., LEDION J., BERTHE L. and FABBRO R. (2000c). Corrosion reactivity of laser-peened steel surfaces. *Materials Engineering and Performance*, **9**, 656–62.

PREVÉY P.S. (1996). Current applications of X-ray diffraction residual stress measurement. *Development in Materials Characterization Technologies*, Van der Voort G. and Friel J. (eds), ASM International, Materials Park, OH, 103–110.

PREVÉY P.S. and CAMMETT J.T. (2002). The effect of shot peening coverage on residual stress, cold work and fatigue in Ni–Cr–Mo low alloy steel. *Proceedings of the 8th International Conference on Shot Peening*, Wiley–VCH, Garmisch-Partenkirchen, Germany.

ROMAIN J.P., COTTET F., HALLOUIN M., FABBRO R., FARAL B. and PEPIN H. (1986). Laser shock experiments at pressures above 100 Mbar. *Physica*, **139–140B**, 595–8.

RUSCHAU J.J., JOHN R., THOMPSON S.R. and NICHOLAS T. (1999). Fatigue crack nucleation and growth rate behavior of laser shock peened titanium. *International Journal of Fatigue*, **21**, S199–209.

SANO Y., MUKAI N., OKAZAKI K. and OBATA M. (1997). Residual stress improvement in metal surface by underwater laser irradiation. *Nuclear Instruments and Methods in Physics Research B*, **121**, 432–6.

SCHERPEREEL X., PEYRE P., FABBRO R., LEDERER G., CELATI N. (1997). Modifications of mechanical and electrochemical properties of stainless steels surfaces by laser shock processing. *Proceedings of SPIE-Europto Conference*, Munich.

SCHIFFNER K. and HELLING C.D.G. (1999). Simulation of residual stresses by shot peening. *Computers & Structures*, **72**, 329–40.

SKEEN C.H. and YORK C.M. (1968). Laser-induced 'blow-off' phenomenon. *Applied Physics Letters*, **12**, 369–71.

SMITH P.R., SHEPARD J., PREVEY III P.S. and CLAUER A.H. (2000). Effect of power density and pulse repetition on laser shock peening of Ti-6Al-4V. *Materials Engineering and Performance*, **9**, 33–77.

SOLLIER A., BERTHE L. and FABBRO R. (2001). Numerical modeling of the transmission of breakdown plasma generated in water during laser shock processing. *European Physical Journal Applied Physics*, **16**, 131–9.

THOMPSON S.D., SEE D.W., LYKINS C.D. and SAMPSON P.G. (1997). Laser shock peening vs shot peening – a damage tolerance investigation. *Surface Performance of Titanium Alloys*, Gregory J.K., Rack H.J. and Eylon D. (eds), 239–51.

VACCARI J.A. (1992). Laser shocking extends fatigue life. *Laser Technology*, Vaccari J.A. (ed), American Machinist, 62–4.

WHITE R.M. (1963). Elastic wave generation by electron bombardment or electromagnetic wave absorption. *Journal of Applied Physics*, **34**, 2123–4.

YANG J.M., HER Y.C., HAN N.L. and CLAUER A.H. (2001). Laser shock peening on fatigue behavior of 2024-T3 Al alloy with fastener holes and stopholes. *Materials Science and Engineering*, **A298**, 296–9.

YANG L.C. (1974). Stress waves generated in thin metallic films by a Q-switched ruby laser. *Journal of Applied Physics*, **45**, 2601–7.

ZHANG Y.K. and CAI L. (1996). Effects of laser-induced shock waves on metals. *Progress in Natural Science*, **6**, S428–31.

ZHANG H. and YU C. (1998). Laser shock processing of 2024-T62 aluminium alloy. *Materials Science and Engineering*, **A257**, 322–7.

ZHANG Y.K., ZHANG S.Y., ZHANG X.R., CAI L., YANG J.C. and REN N.F. (1997). Investigation of the surface qualities of laser shock-processed zones and the effect on fatigue life of aluminum alloy. *Surface and Coatings Technology*, **92**, 104–9.

ZHANG H., ZHANG Y.K. and YU C.Y. (1999). Surface treatment of aluminium alloy by laser shock processing. *Surface Engineering*, **15**, 454–6.

ZIENKIEWICZ O.C. (1977). *The Finite Element Method*, Third Edition, McGraw-Hill Book Company (UK), London.

ZIENKIEWICZ O.C. and TAYLOR R.L. (1991). *The Finite Element Method*, Fourth Edition, McGraw-Hill Book Company (UK), London.

Index